MANIPULATION AND DARK PSYCHOLOGY

THE COMPLETE GUIDE TO MASTER THE ART OF PERSUASION. LEARN SOME USEFUL TIPS ON HOW TO MANIPULATE AND CONTROL THE MIND AND INFLUENCE OTHER PEOPLE.

William Hansen

© Copyright 2020 - All rights reserved.

The content contained within this book may not be reproduced, duplicated or transmitted without direct written permission from the author or the publisher.
Under no circumstances will any blame or legal responsibility be held against the publisher, or author, for any damages, reparation, or monetary loss due to the information contained within this book. Either directly or indirectly.

Legal Notice:
This book is copyright protected. This book is only for personal use. You cannot amend, distribute, sell, use, quote or paraphrase any part, or the content within this book, without the consent of the author or publisher.

Disclaimer Notice:
Please note the information contained within this document is for educational and entertainment purposes only. All effort has been executed to present accurate, up to date, and reliable, complete information. No warranties of any kind are declared or implied. Readers acknowledge that the author is not engaging in the rendering of legal, financial, medical or professional advice. The content within this book has been derived from various sources. Please consult a licensed professional before attempting any techniques outlined in this book.
By reading this document, the reader agrees that under no circumstances is the author responsible for any losses, direct or indirect, which are incurred as a result of the use of information contained within this document, including, but not limited to, — errors, omissions, or inaccuracies.

Table Of Contents

INTRODUCTION 6

CHAPTER 1 THEORETICAL OVERVIEW 10

CHAPTER 2 PRACTICAL AND HISTORICAL OVERVIEW 18

CHAPTER 3 UNDERSTANDING DARK TRIAD PERSONALITIES 26

CHAPTER 4 TYPES OF EMOTIONAL MANIPULATION 34

CHAPTER 5 CATEGORIES OF EMOTIONALLY MANIPULATIVE BEHAVIOR 40

CHAPTER 6 SIGNS THAT YOU'RE BEING MANIPULATED 44

CHAPTER 7 FACTORS THAT MAKE YOU VULNERABLE TO MANIPULATION 50

CHAPTER 8 DARK PERSUASION METHODS 60

CHAPTER 9 HOW TO USE DARK PSYCHOLOGY IN REAL LIFE 68

CHAPTER 10 WHY ANALYZE PEOPLE 76

CHAPTER 11 HYPNOSIS 82

CHAPTER 12 MIND CONTROL 90

CHAPTER 13 BODY LANGUAGE 96

CHAPTER 14 HOW TO RECOGNIZE A HARMFUL RELATIONSHIP 104

CHAPTER 15 BRAINWASHING 112

CHAPTER 16 THE ROLE OF DEFENSE 120

CHAPTER 17 WHAT IS NLP (NEURO LINGUISTIC PROGRAMMING)? A METHOD TO ENHANCE PERSONAL DEVELOPMENT 128

CHAPTER 18 DECEPTION 138

CHAPTER 19 MIND GAMES 146

CHAPTER 20 PERSUASION VS. MANIPULATION 154

CHAPTER 21 MIRRORING 162

CHAPTER 22 SEDUCTION AND DARK PSYCHOLOGY 168

CHAPTER 23 HOW TO PROTECT YOURSELF AGAINST EMOTIONAL PREDATORS 174

CHAPTER 24 SPEED READING PEOPLE 182

CHAPTER 25 THE SECRETS TO TAKING CONTROL OF YOUR LIFE 188

CHAPTER 26 HOW TO AVOID DARK PERSUASION AND STOP BEING ABUSED ... 196

CHAPTER 27 APPLYING MANIPULATION AND MIND REPROGRAMMING IN DIFFERENT ROLES 204

CHAPTER 28 MASTERING YOUR EMOTIONS 212

CONCLUSION 222

Introduction

Dark Psychology is a branch of psychology that fascinates people across the globe who are interested in topics like understanding the criminal mind, better understanding the darker thoughts that control human behavior at all ages, and the conscious actions people take to influence others using psychological manipulation.

At its core, Dark Psychology is the specified study of the more wicked side to human nature: what defines it, how to observe it, where the lines are and how it can be utilized for both constructive and nefarious purposes. It covers mild uses like a clever car salesman who continuously has the best sales numbers on his team because he is able to read his customers and build an amiable connection with them based on observations to the severe uses like studying the mind of criminals who use their understanding of human behavior to victimize others.

Persuasion, manipulation, and other forms of influence are ubiquitous. You can pick up on some obvious signs here and there, but there are also hidden secret ways that others control you which you might never be able to fully comprehend.

Many reasons exist that can make you yearn to be a more persuasive person. Perhaps you feel as though you are already under the deep influence of others and you wish to break yourself free. Maybe you are the kind of individual that can easily fall for the charm of others and now is the moment for you to be able to better protect yourself against any type of influence that might happen to you.

Perhaps you are trying to sell something, maybe yourself or your brand, and you need to figure out how to get people to be more persuaded by you in order to help you achieve the things that you want in this life. No matter where you are or what you are trying to do, you have all the tools that you will ever need to be persuasive or influential with you already.

Before getting into this book, there are a few things that you need to know to be introduced to this topic, in order to get into the right mindset as you read through this text. First, understand that there are no two manipulators that are alike. There are no two easily persuaded people that are the same either. Though it might seem like this sometimes, especially since you can influence a group all at once, you can't let yourself fall into a thinking pattern where you place everyone in the same category.

What you also have to understand is that you should have an open mind with how you interpret the types of manipulation or persuasion that you will see as you read through the book, and afterwards in everyday life. We have tips and tricks to assist you to be more persuasive but look for your own methods as well. Apply the things we are talking about practically in a way that helps your life and with methods that are individual to your experiences and circumstances.

Remember, above all this takes practice. You won't be able to understand the human brain and be a persuasive person overnight. You will certainly be more aware of these kinds of things and the switch will be much quicker than you realize. Don't blame yourself for not being aware of the ways that you have been manipulated in the past. Regret isn't going to do you any good in this journey, so it's best to leave those feelings of, "I wish I would have known this sooner," behind. All that you can do now is move forward, and we are going to help you every step of the way!

Chapter 1
Theoretical Overview

The science of psychology is one that hasn't existed for as long as many other sciences. While geology, biology, and other hard sciences have existed for hundreds or even thousands of years, psychology was born in 1874 when German physiologist Wilhelm Wundt published a book outlining the connections between the body's reactions and the human mind titled Principles of Physiological Psychology.

The idea of dark psychology became popularized when the dark triad of human traits became a widespread topic. This dark triad of personality traits can be found in just about everybody, but the combination of the three traits—psychopathy, narcissism, and Machiavellianism—makes up the most complete idea of a bad person, in their rawest and simplest form. Research of these dark three traits can be found especially in the fields of criminal or deviant psychology. Everyone can show signs of at least one of these traits, but they are a staple of manipulative and malevolent people. When the ideas of these dark psychological qualities became associated with deviance or criminal behavior, more research was done on the ideas behind what made some people deviant, and what made some psychology "dark." These ideas of "dark psychology" ultimately stem from methods of manipulation in order to convince people to lean your way, or to get what you want out of the people around you.

There's also a very important balance that has to be maintained when it comes to dark psychology—it isn't something that you should use whenever you want something out of someone. Dark psychology is something you use when the thing you're trying to get from someone will ultimately help you to have more opportunities—it should be something very important to you or to someone else. If you use these tools simply for the purpose of getting whatever you want, you won't develop as a person for using them, and you'll likely only know how to rely on those tools when you need something in the future. When you need something very important which you can't get in another way, these methods can work very well in a stressful situation or one in which you need a quick solution. However, you shouldn't rely on them if you don't absolutely need to. While mind control and dark psychology are incredibly important parts of psychology and are relevant facets of the human mind, they shouldn't ever be taken lightly, and they aren't things to be messed with when it comes to using them to your advantage. When you really need something from someone, use dark psychology as a way to streamline your goals and your intentions in a way that causes the least amount of damage to other people as possible. While the goals of dark psychology are self-serving and not to service others, you also shouldn't go out of your way to hurt other people or cause them trouble just for the sake of furthering a goal. There are some things that can't be avoided when you're using manipulation tactics, but you should always make it a priority to do as much positive work you can for yourself while minimizing the negative work you do unto others in the process.

That being said, you also shouldn't expect dark psychology to work quickly. When you manipulate someone, it should be a process instead of just a quick fix to a problem. For example, if you're trying to persuade someone to make a certain decision in your favor, it isn't a forceful thing, and persuasion is something that takes more than one encounter to work to its fullest capabilities. While trying to persuade someone, it's ideal to bombard them with subtle but effective ideas that are more likely to get them to understand your side and be sympathetic to your ideas. After a long enough time of that exposure to your ideas and your way of thinking, they'll be much more open to your options and receptive to what you have to say. If you were to just try and force their opinion to change, you probably would've been met with failure as that kind of forceful persuasion can backfire except for specific situations. However, because you took your time in this situation and really hammered in the ideas for sympathy to your cause over a longer period of time, you ended up getting exactly what you wanted because you took the longer and more patient approach to your problem. Dark psychology is sometimes painted as a fast solution to all your problems and the kinks in your relationships. However, just like any method of getting past an issue, you see the best and more effective results when you take your time and patiently observe the process. This applies to all the methods of persuasion and control over others that you'll see and learn — they all work best if you can have the patience to wait them out and let persuasion run its course slowly rather than forcing the process. When we let these things work themselves out more slowly, we let the person we're trying to persuade believe that their decision to support you was ultimately their idea and that they weren't affected by your persuasion much or at all. This also makes slower persuasion much more likely to work when it comes to people who are more determined and independent people who are less likely to agree with you by force simply because they like to feel in control of their own choices. When you persuade them subtly, you offer them the illusion of choice while making the "choice" you want them to make seem much more favorable by comparison.

This is just one of the many ways you can get people to side with you without having to force their hand or use any kind of violence or intimidation. Ideally, you wouldn't have to use any kind of covert persuasion or manipulation in order to achieve your goal, but it's a part of modern life that has become almost inseparable from business and many other spheres of work. In the business world or in any world where communications happen often or agreements have to be reached between different parties who might have entirely different agendas, there's usually some degree of persuasion or manipulation— usually on all sides—in order to reach the most fruitful or pleasant compromise. That doesn't mean that the persuasion is malicious or that the goal of the persuasion is simply to harm the other parties in order to get what you want. Often, persuasion is a normal tool that all humans use as an evolutionary trait that developed when bartering became a more sophisticated action that required a lot of trade. When the way we spoke to other parties became more important than the deal itself, humans developed more of an understanding of what most humans see as most pleasurable, as well as the ability to more accurately read the emotions of people in order to reach the fullest understanding of what each group of people wanted.

Of course, there will always be some aspects of dark psychology which are mainly used in order to bully people into submission or to force the hands of others in order to get what you need out of them. This is never usually preferable, but there are some situations where it's best to go about it in a more forceful way. Additionally, there are some situations where you can't wait for persuasion to be slow and subtle in order to be effective. For some scenarios, you simply need to force what you need out of people. Although it isn't ideal to have to force things out of people this way, it's usually required if you want to reach your full potential in that situation. While you might have ideally been in a situation where you could drop a lot of hints toward someone that they should support you and wait for them to come to that decision on their own terms, you might instead be on a time crunch. Because you have less time, you might be forced to resort to fear tactics and manipulating the person using that fear or intimidation in order to scare them into submission and supporting you. This is one of many situations where even if you would have liked to use more docile or subtle techniques; you're forced by circumstance to resort to potentially more harmful or hurtful tactics of dark psychology. These tactics are simply sometimes needed if you want to have the most effective results out of people. The point of dark psychology is not to befriend the people you're manipulating; after all—the point is to get what you need out of them and little else. In some situations, you might keep that person's connection for future manipulations if they can be involved in your long-term plans.

While the line can sometimes be thin, it's especially important to understand where the boundary falls for you as it pertains to the difference between manipulating someone from a business perspective and being toxic toward that person in a close or intimate way. Once the manipulation becomes personal and the person you're taking advantage of really becomes the victim of your consistent manipulation, you may want to assess why you're treating them that way and if your ends are really justifying the means of that manipulation. While the morals behind dark psychology in a broader sense vary wildly from situation to situation, you shouldn't let yourself become an abuser of someone innocent just for the sake of your own professional or personal gain. If it starts to feel as though you're regularly taking advantage of someone just for the sake of feeling control over them and you've stopped getting anything of proportionate value out of that manipulation, you might be a toxic person—manipulating others not for any real goal which is worth that manipulation, but simply doing so out of the malice you have for them and the desperation you have for social power and control.

You don't have to use dark psychology as a way to have control over others at all times. Usually, it's a tool to fulfill the desires of yours and a way for you to potentially save people time and emotional energy. Although the morality of making a decision with heavy consequences for someone before they can make it themselves is questionable, sometimes you truly do know something that others don't, which allows you to streamline an emotional process or decision which might have otherwise been much more emotionally challenging for everyone involved. For example, if you knew someone who wanted to attend a party or other function for the purpose of meeting someone they already have in mind, you might be able to save them the trouble by preventing them from attending if you know the person they have in mind would be toxic or otherwise problematic for them. If you know this person would be bad for them, you have many ways to use dark psychology in order to prevent them from going to the function—you don't have to use fear or intimidation tactics in order to get them to stay.

Chapter 2
Practical and Historical Overview

There are many hypotheses about the roles and impulses of the human brain. A lot of research also explores how people influence their beliefs, belief systems, and behavior patterns. On the other hand, most hypotheses would address how individuals usually act in certain situations, environments, and contexts.

How do you fully understand the mind of the other person? How do you use the other person's mental awareness to persuade, trick, or manipulate the other person into your thinking? The simplest solution to this problem is to consider what inspires and motivates the other person to take action. Do they do something for the money, or as a passion? Do they do it for glory? Do they do this for authority? Do they do it for the sake of fame? Understand his personality and state of mind. Once you are equipped with the other person's drive and motivations, it is now time to strategically position your discussions and demands so that he and his motives benefit. He will immediately accept your request with little to no opposition. Will you know the reason for this acceptance? Perhaps because he will see you as being very much like him due to the way you have monitored the discussion and made your application. He will feel the inherent duty to fulfill your request as such. This method is the fundamental principle of persuasion, deception, and manipulation.

This research's central theme is to persuade, exploit, and trick the other person without being identified or heard by the other person. In other words, your goal is to hide and hide from your purpose, your attempts, and intentions.

If you ever studied the rules of logic, you would see that manipulating, deceiving, or persuading a person might put you in this rather absurd situation of making a mistake. Fallacies are in themselves absurdities, which means your argument may not always be valid because it defies specific laws.

If you want your point always to be valid and you want to influence the minds of others by the most rational means, then you might find that there are a lot of mistakes in psychology. There are arguments such as argumentum ad baculum, argumentum as edacitam, argumentum ad verecundiam, and others for "fraudulent appeals." Such mistakes are committed by people because, of course, they respond to impulses, authority, or turn of events they cannot foresee.

That is the good part about the real world, though—not all people are convinced that the "visual" claim is all that comes under sound logic. A testable theory means that facts always transcend rational laws in this universe. Their law defiance (as people can see it), however, makes the world much more real. Often people try to predict the weather, which is entirely unpredictable, of course. Some forecasts nail it, and in an unforeseen nature, some will work out.

It is all about empathizing with others and then putting a credible and sound argument into their heads that work well for your benefit. That is because the real world's individuals are more likely to respond to what they believe and not to what the truth in their heads says. There are many reasons why somebody else might want to manipulate, mislead, and persuade. Now, the question is: is your ethical way of thinking manipulating, deceiving, and persuading another person? For the answer, it depends on you. No one can decide if what you are doing is moral or not except you. Nonetheless, there are many social situations where you need to perform specific methods of bribery, deceit, and persuasion to achieve your goals.

Let us assume you are the U.S. president, and Russia is trying to engage you as an illustrative example in the global political arena in a nuclear arms race. Of course, you do not want to start another World War. The Russian leader is, therefore, your individual goal, and you aim to prevent conflict and to preserve peace. In this case, it would be beneficial to use techniques of persuasion, deception, and manipulation to foster your role of world peace and harmony.

For example, the target people are your clients and consumers if you are in the business sector. Your goal is to get them to buy your products or use your services.

If you are an environmental attorney, your target people are politicians, lobbyists, and the public. Furthermore, in protecting the environment, you intend to persuade them to join your cause.

When you look at the topic of mind control closely, it is, undoubtedly, basically a game of persuasion that is practiced every day. It would, of course, be up to you to consider the purpose of learning these tricks. As Machiavelli would say, everything becomes a tool for a particular purpose. Nevertheless, the art of mind-manipulation does not mean that you refuse the use of free will to your goals. Instead, you give them something that they most definitely are searching for–a feeling of a positive choice that acts as a guide to their behavior.

Ultimately, the magnitude of your actions and your system of private beliefs will determine whether or not the tactics and strategies you use are ethical.

If you are an entrepreneur, your customers are your target audience, and your position or purpose is to get them to buy your products or use your resources.

In this regard, one of the most effective ways to change the way you think about your customers (or your goal) is to change your way of speaking. In other words, to create what you offer and what your customers want, you must use the right phrases. If you are watching commercial advertisements on TV and YouTube right now, you will understand that short business clips are attractive because they use the words (or jargon) to address the specific issues of their target people. Therefore, there is an emotional meaning that only the terminology used by the product or service producers knows the point of view, circumstance, or environment.

Today, when you think about how ads show on TV, they all deliver two kinds of sensation-pleasure and pain. Primarily, they draw on the two phenomena that have altered humanity's path forever. Philosophers, religious orders, and tyrants all played with the notion of sensations with a similar belief that everyone would want to do something to enjoy or prevent pain. Millenniums later, with the same idea, people still act. That is why the art of manipulation and mind control is still there.

How do you think others like a coffee brand? The response is a pleasure. It is still the game of joy against pain. Some would assume that they want double shots of espresso to achieve the pleasurable feeling of being awake even quicker because they are frustrated with slowness. I want to get the bitter taste of coffee as soon as I take a sip. Similar to the latte men, though, they prefer milk over coffee, and they want to experience the caffeine as it is in the form of slow waves of the bloodstream. Everyone will appeal to a separate set of people playing to their needs.

Try to tell an espresso addict that he can drink more caffeine than three tall cappuccino glasses in a large glass of latte. More often than not, this espresso lover would try to switch to latte the next morning because he is more likely to get what he wants quickly. At the end of this statement, people are not mostly worried about how to get what they want— they just want a strategy to bring them closer to their goals in a faster manner. We want to believe that all the methods we use are shortcuts to a pleasant experience.

With that in mind, it is rather straightforward for individuals to adjust their belief systems if the move is to ensure achievement. Therefore, on the market out, there are just too many products of the same type. Imagine the number of toothpaste brands. Spread the term, however, that one brand is likely to cause a person's teeth to fall off when he is in his 80s—it almost guaranteed that individuals who are that brand's diehard fans would change or do a rigorous review of their use of that item.

That is why when you are trying to learn how to influence the minds of people around you, always try to know how they feel about a brand and what it is doing to them. You buy a product, one way or the other because you think you want it. But once somebody gives the concept that they shouldn't want it anymore because it will hurt them, they will reject it. However, they would instead look for a similar item, maybe 7 out of 10 times, then stop using it every day. If you can offer cigarettes that do not cause cancer problems and spoil the mouth but still offer the same level of satisfaction as other brands, then you are almost guaranteed to get rich. People do not want cancer, but their cigarettes do not want to say no.

Why is this the case? It is because what they get used to is all they want. Now change that perspective and offer a benefit or easily offer something to make your life better. They are going to risk their lives if they refused to take something. The art of mind control is comparable—the unconscious would always say aims for going towards avoidance of pleasure and pain.

Another fundamental principle for manipulating, deceiving, and persuading a target person is the "authority declaration" strategy. This strategy means that a mere suggestion or claim by a power figure can often alter and modify a person's visual memory to create a distinct memory beneficial to the power figure. Keep in mind that each person's power figure is different. Many individuals will see a university professor as an illustrative authority figure, while some individuals will see a person who has learned from hard knocks as an authority figure. In other words, you must put yourself in a scenario in which you will understand the authority figure of your target person.

You may think that there are people who do not believe in energy. Okay, you are right, so it is not going to be a common tactic to influence people. However, if you look at the history of this civilization, people have always been rooted in any kind of organization. People have still been searching for someone to send them the information they need to improve their lives, or they should look up to them as an inspiration. While individuals may think it is possible to ignore and dismiss the significance of government, they will still be looking for a leader.

You must either have a leader's characteristics or have your words come from a talented person to be able to influence an individual. That way, those listening to the phrases you say wouldn't doubt your words. When people who are admired for their salesmanship can convince others, more people are going to buy the product.

Chapter 3
Understanding Dark Triad Personalities

Dark psychology is not a single, universally applicable medical diagnosis that can be applied across all cases of deviant personalities. There are, in fact, a wide variety of ways that dark psychology may manifest itself in someone's psychological and behavioral makeup. There is no absolute division of one deviant personality type from another, and many deviant personalities with prominent features of dark psychology may display elements of more than one manifestation of dark psychology.

It is important to remember that although the internet has spawned a huge growth in problems resulting from dark psychology, these traits have been part of human culture since ancient times. In fact, one of the dark psychology profiles we will explore, Machiavellianism, takes its name from a medieval politician. Another, narcissism, takes its name from an ancient mythological character. Together, the three dark psychology profiles—psychopathy, Machiavellianism, and narcissism—make up what is known as "the Dark Triad."

Psychopathy

Psychopathy is defined as a mental disorder with several identifying characteristics that include antisocial behavior, amorality, an inability to develop empathy or to establish meaningful personal relationships, extreme egocentricity, and recidivism, with repeated violations resulting from an apparent inability to learn from the consequences of earlier transgressions. Antisocial behavior, in turn, is defined as behavior based upon a goal of violating formal and/or informal rules of social conduct through criminal activity or through acts of personal, private protest, or opposition, all of which is directed against other individuals or society in general.

Egocentricity behavior is when the offending person sees himself or herself as the central focus of the world, or at least of all dominant social and political activity. Empathy is the ability to view and understand events, thoughts, emotions, and beliefs from the perspective of others, and is considered one of the most important psychological components for establishing successful, ongoing relationships.

Amorality is entirely different from immorality. An immoral act is an act that violates established moral codes. A person who is immoral can be confronted with his or her actions with the expectation that he or she will recognize that his or her actions are offensive form a moral, if not a legal, standpoint. Amorality, on the other hand, represents a psychology that does not recognize that any moral codes exist, or if they do, that they have no value in determining whether or not to act in one way or another.

Thus, someone displaying psychopathy may commit horrendous acts that cause tremendous psychological and physical trauma and not ever understand that what he or she has done is wrong. Worse still, those who display signs of psychopathy usually worsen over time because they are unable to make the connection between the problems in their lives and in the lives of those in the world around them and their own harmful and destructive actions.

Machiavellianism

Strictly defined, Machiavellianism is the political philosophy of Niccolò Machiavelli, who lived from 1469 until 1527 in Italy. In contemporary society, Machiavellianism is a term used to describe the popular understanding of people who are perceived as displaying very high political or professional ambitions. In psychology, however, the Machiavellianism scale is used to measure the degree to which people with deviant personalities display manipulative behavior.

Machiavelli wrote The Prince, a political treatise in which he stated that sincerity, honesty, and other virtues were certainly admirable qualities, but that in politics, the capacity to engage in deceit, treachery, and other forms of criminal behavior were acceptable if there were no other means of achieving political aims to protect one's interests.

Popular misconceptions reduce this entire philosophy to the view that "the end justifies the means." To be fair, Machiavelli himself insisted that the more important part of this equation was ensuring that the end itself must first be justified. Furthermore, it is better to achieve such ends using means devoid of treachery whenever possible because there is less risk to the interests of the actor.

Thus, seeking the most effective means of achieving a political end may not necessarily lead to the most treacherous. In addition, not all political ends that have been justified as worth pursuing must be pursued. In many cases, the mere threat that a certain course of action may be pursued may be enough to achieve that end. In some cases, the treachery may be as mild as making a credible threat to take action that is not really even intended.

In contemporary society, many people overlook the fact that Machiavellianism is part of the "Dark Triad" of dark psychology and tacitly approve of the deviant behavior of political and business leaders who are able to amass great power or wealth. However, as a psychological disorder, Machiavellianism is entirely different from a chosen path to political power.

The person displaying Machiavellian personality traits does not consider whether his or her actions are the most effective means to achieving his or her goals, whether there are alternatives that do not involve deceit or treachery, or even whether the ultimate result of his or her actions is worth achieving. The Machiavellian personality is not evidence of a strategic or calculating mind attempting to achieve a worthwhile objective in a contentious environment. Instead, it is always on, whether the situation calls for a cold, calculating, and manipulative approach or not.

For example, we have all called in sick to work when we really just wanted a day off. But for most of us, such conduct is not how we behave normally, and after such acts of dishonesty, many of us feel guilty. Those who display a high degree of Machiavellianism would not just lie when they want a day off; they see lying and dishonesty as the only way to conduct themselves in all situations, regardless of whether doing so results in any benefit.

What's more, because of the degree of social acceptance and tacit approval granted to Machiavellian personalities who successfully attain political power, their presence in society does not receive the kind of negative attention accorded to the other two members of the Dark Triad—psychopathy and narcissism.

Narcissism

The term "narcissism" originates from an ancient Greek myth about Narcissus, a young man who saw his reflection in a pool of water and fell in love with the image of himself. In clinical psychology, narcissism as an illness was introduced by Sigmund Freud and has continually been included in official diagnostic manuals as a description of a specific type of psychiatric personality disorder.

In psychology, narcissism is defined as a condition characterized by an exaggerated sense of importance, an excessive need for attention, a lack of empathy, and, as a result, dysfunctional relationships. Commonly, narcissists may outwardly display an extremely high level of confidence, but this façade usually hides a very fragile ego and a high degree of sensitivity to criticism. There is often a large gulf between a narcissist's highly favorable view of himself or herself, the resulting expectation that others should extend to him or her favors and special treatment, and the disappointment when the results are quite negative or otherwise different. These problems can affect all areas of the narcissist's life, including personal relationships, professional relationships, and financial matters.

As part of the Dark Triad, those who exhibit traits resulting from Narcissistic Personality Disorder (NPD) may engage in relationships characterized by a lack of empathy. For example, a narcissist may demand constant comments, attention, and admiration from his or her partner, but will often appear unable or unwilling to reciprocate by displaying concern or responding to the concerns, thoughts, and feelings of his or her partner.

Narcissists also display a sense of entitlement and expect excessive reward and recognition, but usually without ever having accomplished or achieved anything that would justify such feelings. There is also a tendency toward excessive criticism of those around him or her, combined with heightened sensitivity when even the slightest amount of criticism is directed at him or her.

Thus, while narcissism in popular culture is often used as a pejorative term and an insult aimed at people like actors, models, and other celebrities who display high degrees of self-love and satisfaction, NPD is actually a psychological term that is quite distinct from merely having high self-esteem. The key to understanding this aspect of dark psychology is that the narcissist's image of himself or herself is often completely and entirely idealized, grandiose, and inflated and cannot be justified with any factual, meaningful accomplishments or capacities that may make such claims believable. As a result of this discord between expectation and reality, the demanding, manipulative, inconsiderate, self-centered, and arrogant behavior of the narcissist can cause problems not only for himself or herself, but for all of the people in his or her life.

The Dark Triad in Practice

The professional workplace has acknowledged the presence of people exhibiting Dark Triad characteristics. The following diagram illustrates that they are tolerated for their efficiency and their ability to get things done but contrasts that ability with the negative effects it has on their ability to form personal relationships:

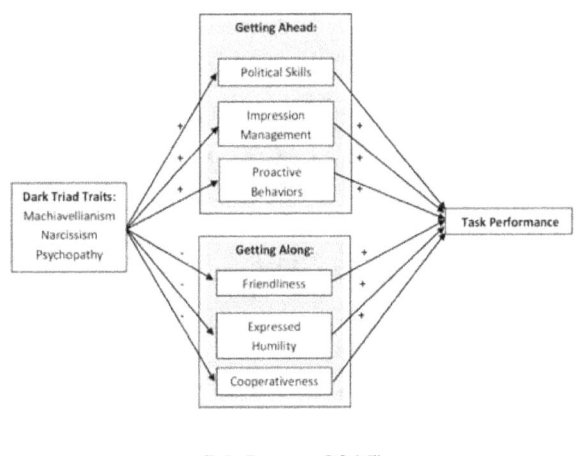

(McLarty, 2015)

The remainder of this discusses a wide variety of people and situations in which you may find one, two, all three, or some combination of these Dark Triad personalities working in concert around you.

The clinical descriptions are easy enough to categorize, and in isolation, it can be fairly straightforward to separate one type of dark psychology from another. The real world is a lot messier. Many of us have grown accustomed to so-called "toxic relationships," whether they are relationships with our partners, our co-workers, our family members, our bosses, or our political and community leaders. In addition, manifestations of dark psychology are often far more mundane than the dramatic examples we see in major television and film productions about the romantic lives of serial killers and other criminals. The more we accept these relationships as normal, the more difficult it will be to identify them as problematic.

Remember that psychological, emotional, and social predators do not think of themselves as sick. Their lack of morality and empathy, and their adaption from a very early age to live according to rules and methods you may find horribly wrong, can make their presence intimidating. However, you should also remember that even when their amorality and lack of empathy may allow them to enjoy an unjust advantage in relationships, their mental capacities are the result of underdevelopment, not a higher evolutionary state.

Chapter 4
Types of Emotional Manipulation

Some people are always lucky in that they can get what they want at any time the need arises. Sometimes they do so at the expense of others, this, however, is achieved via access to the emotional bank. They can influence your thoughts or emotions to their advantage and leave you a victim and vulnerable.

Owning space

The main aim of emotional manipulation is to make you lose control of your emotions. It will involve making you stagger with emotions which on the other hand will make you even more vulnerable. Advantage will be taken, and they will access you and get all they wanted from you. If there existed a lock to the emotions, I am sure everybody would have their emotions locked away and unlocked only to intimate relationships or where your emotions will be valued. To make sure you are off the steering with your emotions, manipulators will invite you to a place where they know it is new to you but familiar to them. This will keep you off balance, the new environment will give him or her the dominance and feeling of being in control. You are new to the place and the manipulator will take advantage of the window between adaptability and regaining control.

Your words against you

How you talk or react speaks volumes and emotions can be passed along. Manipulators like a talkative person since it is easier to access them due to the link provided; speaking out. If you are the introvert type or a conservative person, it takes more effort to make you open up. Introverts would require tailored questions that will be well planned and will give you away from one by one. The manipulator makes sure the questions are aimed at the emotional state. Personal questions will open you up and you will start speaking with feelings, this is an indicator that manipulation is taking place and it is working. By asking simple and tailored questions that mostly are personal or involve something we like hobbies, interests among others will lead to saturation with emotions. A master manipulator will take advantage of the situation and make us of the questions to establish your beliefs, strengths, and weaknesses without you realizing it.

Guilt

Kind-hearted victims are easily vulnerable to emotional manipulation. Guilt will be used against you, especially if you are so sensitive you may end up giving in to their demands. Guilt will either make you give in or feel bad about yourself. For instance, you may both agree on something and when the time comes to complete the deal the manipulators will pretend to forget or even act as victims of your actions. By doing so they will be finding your soft spot and once they find it will be the target of manipulation. Guilt and sympathy will be served to you, if you are not strong enough you will fall for the play.

They will influence you through that guilt since you will be under their spell and since you now believe they are the victims you will do all they ask just to make sure your 'victims' do not suffer anymore.

Positive and negative emotions

Emotions of sadness or happiness can also be a pathway for emotional manipulation. An emotional manipulator will play with your psychology, he or she will show you that what you might be going through is nothing compared to what they have going on in their lives. By doing this, they try to exalt you and win your trust. If you fall for that and believe there are more needy people than you in the world then you will loosen up and think that you are selfish. You will no longer focus on your big problem, rather you will focus on their 'big unfortunate events' since you will now feel pitiful. Once you trust them, then you give them a key to your emotional bank and surely, they will use it against you. Once you trust them, you might end up offering yourself to assist them, that, however, was their plan from the start; they will have attained their goal.

Anger

Anger is another emotion that can be used to induce emotional manipulation. Some people are natural peacemakers, they avoid confrontations and conflicts in all ways possible. Once a manipulator realizes you are this type of person, he or she will use anger, aggressive language or raise his or her voice or even drop several threats. These aggressive techniques are tailored just to make tick. The secret behind this aggressive approach is to induce fear and discomfort so that you can give in hastily without taking a second to think through. Once you give in to their demands, they now get control over you and now can manipulate you in whatever direction or way that pleases them. They use this opportunity to get what they wanted from you since you will be cooperative earing to bring another instance of acute aggression.

Self-discipline and confidence

Being self-driven and confident is a very strong barrier to the effects of emotional manipulation. With the right mindset, you become less vulnerable to emotional manipulation attacks. Insecure and sensitive people are the easiest target for emotional manipulators. They are easily spotted and accessible, they put their needs behind those of others and are often feeling the need to please. All a manipulator needs are to be caring, sensitive and with an urge to help out. The needy part of sensitive people exposes them, and the emotional manipulator will see it as a gate pass to influencing your thoughts, perceptions, and feelings to his or her advantage. With time the emotions break open and they are exploited easily since the manipulator was disguised as a caring and sensitive person. As the saying goes, birds of a feather flock together; the feeling of sharing the same trait will open them up for manipulation without their knowledge.

Surprises

Negative surprises are also another mechanism used to keep people off balance. When bombarded with the new unexpected news that comes with a limited timeframe will lead to panic. As you panic, you get little time or none to think of a counter move. They may be good enough to trap you with suggestions as they pretend to help yet it is a plan made to make you unstable both psychologically and emotionally. Once you become unstable and overwhelmed by the sudden change of events it becomes their opportunity to influence your decisions and any other emotion, they are interested in. They may even consider making more moves that will bind your relationship with him, her or them so that they can utilize that window of opportunity created by the panic moment. You may not realize it since they appear to be assisting whereas they are using you for their benefits.

Criticism

Criticism is also a tool for emotional manipulation. The manipulator will say bad things of you, ridicule you or even dismiss you. He or she will make sure her mission of dismantling you succeeds. Once you have had enough you end up off balance and believe they are much superior compared to the inferior you. You will feel so down, and their opinions will stick. Once you are in this state you are vulnerable. The manipulator will make sure you understand that you can never be good at anything no matter what you do or invest in. This will get into you and you will be emotionally distressed. You will feel hurt and not worthy of anyone's help. They will then pretend to have answers to your problems. He or she will give you tips and suggestions that are so genuine looking and constructive. Once these well-outlined answers transform you and get you out of it, they threw you will worship them.

Once they have your attention, they can then make you do what they want or influence you.

Doubt

Doubt and uncertainty are also other forms of leverage in emotional manipulation. You will receive a silent treatment until you start doubting your actions or words that you may have used the last time. The manipulators will do this deliberately to stir up the feeling of doubt, once you give in and break the silence by acting as the cause of the silence treatment will be a good chance to be taken advantage of. This creates a window of opportunity and they will manipulate you.

Ignorance

Pretending to be ignorant of your duties will also get things done. You may want to do something, but you want it done by someone else, let us say your spouse. She or he will note something is off and will try to make it right, but you pretend to be good with it but since they know it has to be right, they will do it anyway.

Chapter 5

Categories of Emotionally Manipulative Behavior

Manipulation in relationships

Long-term control can have serious effects on close relationships, consisting of those between good friends, family members, and romantic partners. An adjustment can deteriorate the health of a relationship and lead to the poor psychological health of those in the relationship or even the dissolution of the relationship.

In a marriage or collaboration, manipulation can trigger one partner to feel bullied, isolated, or useless. Even in healthy relationships, one partner may accidentally control the other in order to avoid conflict or perhaps in an effort to keep their partner from feeling burdened. Many individuals might even know they are being controlled in their relationship and choose to neglect or downplay it. Manipulation in intimate relationships can take many types, consisting of exaggeration, guilt, gift-giving or selectively demonstrating affection, secret-keeping, and passive aggression. Moms and dads who manipulate their kids may set their children up for guilt, depression, anxiety, eating concerns, and other psychological health conditions. One study also revealed that mothers and fathers who regularly use manipulation techniques on their kids might increase the likelihood their kids will also use manipulative behavior. Indications of control in the parent-child relationship may consist of making the kid feel guilty, lack of accountability from a mom and dad, minimizing a child's accomplishments, and an important need to be included with tons of elements of the child's life.

Emotional manipulation, therefore, involves access to someone and influencing them for information or any other favors through their emotions. It sucks and it is unethical once you realize what has been done to you. Do not get confused, there are two sides to emotional manipulation. There is an ethical and unethical side. What feels like betrayal is the unethical one. However, mastering emotional intelligence plays a barrier role to these manipulations and keeps you in a safe spot.

People might also feel controlled if they belong to a relationship that has become hazardous. In manipulative relationships, one person might be using the other to meet their own needs at the expense of their friends. A manipulative friend may use guilt or browbeating to extract favors, just like loaning cash, or they might only connect to that good friend when they really need their own psychological needs met and may look for excuses when their friend requires in the relationship.

How to protect yourself against emotional predators

Though manipulation does not cause any harm or put the subject in any immediate danger, it is designed to deceive and change the attitude, reasoning and understanding of the intended subject regarding a particular situation or topic and is good to protect yourself and your loved ones against it.

Social influence, such as a teenager inducted into a culture or a society to interact with different people either at home or at work, is admirable. Any social influence that regards the privilege and right of individuals to decide, without been intimidated, is usually seen as something that is helpful.

Then again, social influence is despised when people beguile others to maneuver their way against other people's will. The impact can be very destructive and generally looked down upon as very weak in nature.

Chapter 6
Signs That You're Being Manipulated

The type of people that the manipulative people target is low self-esteem people, no boundaries people, and desperate people. Now how do you know if you are being manipulated? If you feel like you are constantly criticized and he makes you feel inadequate, then it means you are being manipulated. If you get into an argument and he's giving you the silent treatment, you are being manipulated. If somebody gives you the silent treatment, which makes you go crazy, you start thinking of every scenario, and you start making assumptions because no dialogue is happening that can reassure or bring clarity to your thoughts, then that is a huge form of manipulation.

Ghosting You

And guys do this a lot, even if it is something as simple as ghosting you. Because it trains you to not get used to hearing from him certain times, and you always have to reach out to see how he is doing and checking out to see if he remembers the date that he set for you. They use their profession or their education to delay you finding out the truth or make you feel like they're always the right one. For instance, if you get into an argument with someone and you are dating a lawyer, they will tell you something like, "I've been a lawyer for five years, and

I know what I'm talking about and people that did what you just did need not to be trusted." They sort of use their title to rain over you and make it look like they are the right ones. What happens is that you silently agree to what they are saying because they do know what they are talking about. Because they know when people lie, and they know it through body language.

Demonize your reactions

They tend to demonize your reactions because anytime someone that is manipulating you and they don't want you to be able to express yourself or control the situation, they're going to make you feel like you are the bad guy for reacting the way that you did to the situation. They will flip the script on you because you didn't agree with their actions.

You might tell him, "Hey, babe, I don't know why you just liked this girl's picture on Instagram. I thought that we agreed that you are not going to do this. You show me their stuff. Because it makes me feel embarrassed that my boy is licking the girl's photos and commenting on rubbish on Instagram, and it makes me feel insecure because you are my boyfriend". Then he will say something like, "you are so insecure it's just Instagram, I can't believe that you are seriously talking to me about a comment that I wrote to a girl. First of all, I don't even know her, and she looks nice. Other guys are commenting on her photos, but why do you care because I'm with you". So, they demonize you and make you feel like the way you feel is not accounted for. They make you feel like the way you feel is invalidated because he doesn't know that girl, and maybe he may even be trying to learn that girl.

Using pity

One of the greatest forms of manipulation is by using pity. Because getting pity out of anybody is going to guilt-trip them, so that they feel bad for you and do what you say and hear you out and like whatever trash you want to slip by because they are feeling bad for you.

For instance, if you say, "I just realized that when we were in the get-together, you were nagging to really hanging out with me. You were just doing your own thing. I don't know everybody there, I felt alone, and I understand that you know everybody, but I didn't feel included. Then he will say something like, "Honey, I'm really sorry that you didn't feel included. However, what do you expect me to do, all the people were people that I grew up with. So, I'm sorry that I wasn't holding your hand the entire time. But I did introduce you to some people. You know that I wouldn't do that to you. You know that I am not like that. I was just caught up. Plus, I saw one of my girls from high school and we just started talking. Come on, if you really know me, you know that I wouldn't do something like that. I am not like that."

So, they tried to play on your emotions so that you will think that they are helping you. To think something like, oh, "I do know him. I'm not sure that anyone will want to invite me somewhere and then drop me off and not even associate with me at all or leave me alone". So, you feel bad for yelling at him because it gets overwhelming when you are hanging out with so many people that you haven't seen in a long time.

If you are bothered by the fact that he left you alone, then it means that he left you alone for so long that it became so uncomfortable. It's not a big deal if he's going to leave you for some minutes and go to say hi to someone, but he should introduce you to those people because you guys are in a relationship. So, the reason why you are feeling how you are feeling is that something was wrong.

So, the best way to combine this is to minimize their actions so that you get to stick around. If he says, "I can't believe you would do something like that. You should say, "what do you mean. You do dumb things all the time". These people that always want to downplay what it is that they are doing so that you will feel stupid and feel like you're overreacting on what the offense was.

For instance, let's say that you want to surprise him and leave something cute in his mailbox. So, you drive by his house, and you see another car parked in his driveway, and then you notice that another girl is in his house. And then you think that maybe that is one of his guy friend cars so you drive in his car and instead of you to give him a little bit surprised, you get out of the car because you don't know whose car you are seeing. You knock on the door and then he opens the door halfway and starts asking you what things like, "What are you doing here." And then you answer him, "I'm checking in. Are you well? I noticed that there is an extra car in the driveway, and it's not mine. So why are you not letting me in". and he says, "that is one of my homegirls from high school we haven't talked in a long time, and she just wanted to drop by and catch up."

So, you should say, "why is your friend in your house alone and you didn't even mention it to me. I've never seen this girl in my life, and I never knew that this is one of your homegirls. Why am I just finding out about this"? And he says, "calm down you're just a little extra obnoxious, she's just a friend. She just dropped by to say hi. I didn't even think about mentioning it to you, because it's not about what you think. Because if it was like that, you would just tell me," then he's trying to play ignore and because he wants you to feel guilty. He wants you to feel like how can he cheat on you when in broad daylight when he knows that you can come and visit his house. The best way that this guy used to hide things is in plain sight eyesight because it's so unbelievable.

Glaring and Unbelievable things

They do glare and unbelievable things, and then they try to convince you that what you saw wasn't true. And what you saw couldn't be what you possibly think. It is because it doesn't look like you will do anything like that, and he will have to be a real idiot to do something like that to you. He wants to minimize his action and play ignorant like he has no idea what it is, and you are tripping, and both of them are just friends. He also tries to make rude remarks in the name of humor.

It's so important that it's it in your subconscious mind, and whenever you guys are in an argument or in a situation where you feel intimidated, and you're someone that is easily intimidated by other beautiful girls, then what he says becomes your inner voice. So, the joke that he makes about how big your nose gets into your mind. Because you are thinking about the waitress and it's looking like he's flirting with her because she has a nose that he actually likes or he always makes fun of your crooked tooth, and you are very subconscious about that, and the girl over there has straight teeth.

So, you must pay attention to things like that in the relationship and in friendships because there is always some sort of truth to those little remarks. There is always some sort of underlying truth if somebody is constantly attacking something about you like your physical appearance or playing on your weaknesses because they know that it is going to get you inevitably. But, remove any responsibility or accountability for what they are saying even though they're trying to make it look like a joke.

Chapter 7
Factors That Make You Vulnerable To Manipulation

What is Manipulation? When someone talks of manipulation a lot of ideas or what it means comes to mind. According to the Cambridge Dictionary, manipulation is the process of controlling or influencing someone or something to your own merit, which is always unfair. Many people around the world believe the action of manipulation is wrong. It's also known that something which has its disadvantages also has its advantages, it all depends on how you play your cards. There are various manipulative techniques employed by people to make you do what they want. Toxic people can use them to gain an advantage in relationships. Below are clinical analyses of the various manipulative techniques designed to trap people in relationships and also emotions.

Guilt and Sympathy

Guilt and Sympathy can also be categorized as Emotional Blackmail. It's the most common technique used in personal relationships and plenty of time goes unnoticed. People are forced to doing things or favors for a partner as he bargains with your feelings for him. This is because it always works. This kind of manipulation can go on and on until the victim decides she has had enough. Emotional blackmail leads to partners having unhealthy relationships. During arguments, the partner uses threats or fear to have his or her way. Statements such as "if you leave me, I would rather commit suicide than let you be with someone else", are a norm in the relationship. Such words make you feel guilty about what will happen if you decide to go your own way.

Sympathy and Guilt are sometimes used to exploit other people's emotions who are not in a relationship. The act of doing well always outweighs our emotions. As humans, we always feel bad for people who are in complex situations. We come together to aid someone who is in need. Many feel contented after carrying out humanitarian work.

In such scenarios, you need to ask yourself some serious questions and think deeply about your life. Being in a relationship because someone threatened to kill himself is not viable. When such statements are uttered do not fall for it. It's often a manipulation tactic and a cold threat by the individual. You should know where to draw the line. Be selfish for once and think about yourself.

Criticism

This is whereby the manipulator uses tactics such as belittling, dismissing and ridiculing you. This keeps someone off-balance. Negative criticism is directed at a partner that makes them feel unworthy. The act of criticizing enables them to gain control over you. The manipulator creates a narrative that there is always something wrong with you and you not good enough. This kind of narrative makes you doubt yourself often of what you feel and know. It reaches a point where you do not trust yourself.

Manipulators use your vulnerabilities against you. Most often they consciously create relationships with partners who are insecure. They make you believe they are always right and you are wrong. Continuous criticism by the manipulator makes them gain superiority over you. Their criticism is always negative and they tend to make it sound as if they have your best interest at heart. Giving positive feedback where due is rare to come by.

When you are feeling you not yourself in a relationship that is a red flag. In a healthy relationship someone should be free to express oneself and not be having feelings of marginalization. In the event of realizing this, take appropriate action of walking out of that relationship. Never looking back on it.

Using Flattery, Kindness, and Charm

The use of kindness, charm, and flattery is often more damaging than realized. The generous deployment of these techniques is known as a passive-aggressive type of behavior. The manipulator uses tactics such as gifting someone items, massaging their egos with flattery and lots of compliments. One is left to question the real reason behind the compliments, expensive gifts and paying lots of attention to the victim. These are acts done with ulterior motives, especially when they realize you are about to catch up on the manipulative state.

Bribery is also another technique of manipulating someone. It yields results when implemented. Bribery is a better technique than blackmailing an individual. You can even entice a friend with a better offer. The reward doesn't have to be of monetary value. It can be something that you would have done anyway or had already done prior. Perhaps, in the process of asking for help in your studies from a colleague, you can decide to offer her a lunch treat after studying in exchange.

Establish what the individual truly needs. Once you have established the person's need, then you should try and give it to the person. If one of your friends has a crush on one of your school mates, then you should promise your friend that you will get the school mate's number.

While using flattery or kindness, ensure that you do not make it obvious. Just ensure that you make it appear as if you are determined to do something nice for the friend.

When you have spotted the manipulation aspect, do not accept their gifts or give in to their compliments. These are not sincere kindness but methods deployed to lure you back in. remember they are only carrying out these acts so as to "get" the conditional love which is not there.

Gaslighting

Gaslighting is a manipulation technique that leaves the victim question his state of mind. It's a very insidious technique deployed by manipulators. A most common question to arise through Gaslighting is; "Are you normal?" the partner is always refuting claims of uttering such words or even doing what you are accusing him of, leaving out information and twisting words to his favor. The manipulators also try to push the narrative of losing your mind. The manipulator can talk to you in a way that portrays he knows what best for you. The constant questioning of their knowledge and the act of undermining instincts.

When this act is done continuously, it eats the ability of one trusting her brain to know if she is being mistreated by her partner. The victim strongly believes that she needs the help of the manipulator to keep her on point. A victim can, however, overcome this technique if she keeps a journal of activities unfolding and even sharing with friends.

Lying

Lying is normally used by con artists. Manipulators lying practically in everything they see, hear or know. They create a bunch of lies that are so complexed that they tend to wrap people they cannot differentiate between reality and fake life. These lies can only be disputed by checking for inconsistency in the stories. When the deal is too good think twice before accepting an offer. Abusers use this manipulation technique as they do not have fears about it. Lying can be caught if the victim decides to a background check on the sources of information.

Having the Home Court Advantage

The process of manipulating an individual all depends on the level of control one has over the other. A manipulator will insist on meeting and having your interactions in an area he is comfortable with. This technique is used to gain an edge of control over someone by taking them out of their element. The places where you normally spend it is usually his spots and not yours. These places include his office, house, vehicle and hanging out joints. Such areas make the manipulator exercise his dominance. Someone is easy to control when he is uneasy with the environment.

This technique can be done away with the victim makes it clear that meeting places are where they are both comfortable with. Restaurants are chosen by both of you and events to attend. This leads to a joint partnership between the two of you. Breeding a healthy relationship in the process.

Caressing the Ego

Caressing the ego is most common in personal relationships. The manipulator caresses the ego of the victim by feeding her lies every time. The ego grows with time and in the end, the manipulator has you on his leash. This technique can be avoided if the victim of the act possesses humility.

Making Unusual Request before Your Real Request

This kind of manipulation technique is a slight "mind game" oriented. The tactic involves asking an unusual request, mostly of a higher degree. This throws the person off balance as he wasn't expecting such a request. The manipulator knew how that person was likely to respond, he would have asked for the usual request such as money, favor, shoes. The victim was more likely to respond no. This is because people's minds have been conditioned to avoid these tasks.

For instance, a salesperson knows approaching someone on the street and asking him to buy the items being sold would likely lead to no sale. The salesman would first ask the potential client to do something for such as help him directions. This will build a "relationship" with the person and makes it less likely to turn you down in your presentation.

Voice Raising and Irate Outbursts

Manipulators raise their voices during arguments to intimidate someone. They believed this process of raising the voice aggressively or loudly can enable them to achieve what they wanted. In a relationship, passion can come out in different forms such as tenderness, cute smiles you give one another, laughter, and the desire to share warmth in the arms of another. Passion, however, should not be mistaken by angry unpredictable outbursts during disagreements. In marriages or couples tend to disagree with one another. It is not uncommon. There are various ways of handling conflict, having healthy communication with your partner is one such scenario, not screaming or temper tantrums. The aggressive voice is sometimes accompanied by strong body language.

The manipulator can also start to pick quarrels of non-issues. When this starts to occur, it's a display telling you there are items needed to be ironed out. There are forces at play. In a relationship, someone who's picking random quarrels is either cheating on the partner or looking to end the relationship.

Some manipulators tend to appear calm, cool and collected when the couple is around friends and family. The moment family members have gone the angry outbursts and raising of the voice are witnessed. Even if a partner is on the wrong, one needs to disagree without showing violent tendencies. Provide a constructive criticism or politely talk to the partner. The line should be drawn between passionate and being abusive. While their certain types of manipulators who when a negative event happens, a conflict arises, or things seem to be in chaos, the person becomes super calm. This can also be a manipulation technique. It creates an impression of overreacting by the victim, especially by making you feel like you can't trust your own emotional reactions.

Manipulators can control their partner's emotional responses by doing so. The victim expected some kind of response which she is unable to get from the manipulator. He determines when a situation warrants an emotional response from the intended victim or it's just thought of as being dramatic. The manipulator shows calmness, coolness and collected than was expected.

Inspiring Fear over a Partner then Relief

This technique of manipulation tends to have a high rate of success over its victims. It normally involves making someone fear the worst is going to happen. Creating stories or giving out the wrong information. If the story does not happen that person is likely to be relieved, and then he would be happy enough to grant you whatever you want. It's all about playing mind games with the victim.

Chapter 8
Dark Persuasion Methods

Human beings as a social being are in constant communication for many reasons, i.e. giving information, getting information, asking for help, making promises, telling your feelings and thoughts, or trying to learn someone else's feelings and thoughts, and so on. Communication is established within a certain structure and order.

The concept of persuasion is defined in the dictionary as follows: Convincing, compelling; deceit; "Based on this definition, it will not be wrong to consider persuasion as a form of communication that is realized to achieve the desired aims."

Indeed, when we look carefully, it can be seen that the difference between daily communication and persuasion is to achieve the desired goal. Not every communication phenomenon that is established in daily life is intended for persuasion. Asking about someone's memory only aims to learn about the person's condition and health. However, rather than persuading a person on a particular issue, it should be dealt with to uncover the desired change in the person who is exposed in the final analysis, which should be established with a certain systematic structure.

In the meantime, an important issue should be included here. It is also the effects of communication and how they occur. The effects of communication are:

• Change in the recipient's level of knowledge

• Changes in the attitude of the recipient

• A change in the receiver's open behavior.

In the second stage, the attitude change that came into the agenda is also realized in three ways:

• Strengthening or strengthening the existing attitude

• Change of existing attitude

• New attitude formation

The effects of communication are often expected to occur sequentially and usually do. It is possible to see the effect of communication to a large extent in the change that may occur in open behavior. This is where the difference between daily communication and persuasion comes up.

Persuasive communication is the expected and desired changes in attitude and open behavioral changes that will occur after the information is given. The attitude change expected to occur is determined by some attitude measurement techniques (Likert scale, etc.) developed in cases where open behavioral change can't be observed clearly or if it's not possible for different reasons, for example, an individual's Facebook, and so on.

If it is desired to learn the attitude towards social media, a questionnaire consisting of expressions reflecting this attitude can be prepared. These statements; it allows people to share, enjoy the time, etc. can. It is possible to say that a Likert-type scale was used to measure the attitudes of the respondents to measure attitudes.

The concept and process of persuasion is a subject that has been studied intensively. In general, the biggest factors contributing to the success and failure of communication emerge as convincing communication and its proper structuring. With good understanding and knowledge of persuasive techniques; an educator, an advertiser, or a politician, in other words, it is possible to evaluate anyone whose purpose is to change the thoughts and actions of others.

It should not be ignored that some essential variables exist in persuasion. Each of the variables in persuasion must be identifiable, distinguishable, and measurable.

Scientists working in this field show these variables fall under two headings. These are called "dependent variables" and "independent variables. Arguments are made or occur with the communication process. We know what these variables will be, how they will be formed, and predict and produce their effects.

Dependent variables, on the other hand, have to be done, and convincingly. We often hope to replace dependent variables with independent variables that we manage and control. Dependent and independent variables are called a convincing communication matrix.

The convincing communication matrix is a precise and complete data about all dependent and independent variables in human relationships throughout human life. Independent variables should be considered in many aspects and aspects of communication. However, dependent variables occur only when a person receives a persuasive message in terms of the information process.

Persuasion Techniques

The basis of persuasion is to direct the other person to the thought you desire and to make it normal in the basic belief and vision system. To simplify, it is to make the other person think the way you want. That's exactly what it means to convince. If the other person thinks the way you want, you can take the action that you want to take, that is, buying a product or consuming a product.

Located below are techniques to persuade and convince some of the most effective techniques effectively. Persuasion techniques are not limited to these, but they are important for efficiency. You may encounter many other techniques of persuasion, such as rewarding, punishing, creating a positive or negative perception.

1. Creating Needs

One of the best methods of persuasion is to create a need or to reassure an old need. This question of need is related to self-protection and compatibility with basic emotions such as love. This technique is one of the biggest trumps of marketers in particular. They try to sell their products or services using this technique. The kind of approaches that express the purchase of a product to make one feel safe or loving is part of the need-building technique.

2. Touching Social Needs

The basis of the technique of touching social needs are factors such as being popular, having prestige, or having the same status as others. The advertisements on television are the ideal examples. People who buy the products in these advertisements think they will be like the person in the advertisement or they will be as prestigious. The main reason why persuasion techniques such as touching social needs are effective is related to television advertising. Many people watch television for at least 1-2 hours a day and encounter these advertisements.

3. Use of Meaningful and Positive Words

Sometimes it is necessary to use magic words to be convincing. These magic words are meaningful and positive words. Advertisers know these positive and meaningful words intimately. It is very important for them to be able to use them. The words "New," "Renewed," "All Natural," "Most Effective" are the most appropriate examples of these magic words. Using these words, advertisers try to promote their products and thus make the advertisements more convincing for the liking of the products.

4. Use of Foot Technique

This technique is frequently used in the context of persuasion techniques. The processing way is quite simple. You make a person do something very small first because you think they can't refuse it. Once the other person has done so, you will try to get him to do more, provided he is consistent within himself.

First, you sell a product to a person at a very low price. Then you get him to buy a product at higher prices. In the first step, you attract him to yourself, so you convince him to buy it. In the second step, you convince yourself to buy products at a higher price. Their acceptance of a small thing will help you to fulfill the next big demand from you.

After refusing the small request from the other party, you feel a duty to make a big request from the same person. This is usually the case in human relations. For example, you agree when your neighbor comes and asks you if you can keep an eye on the shop for a few hours. If your neighbor comes to ask you to look at the shop all day, you will feel responsible and probably accept it. This means that the technique of putting a foot on the door is successfully applied.

5. Use of Orientation from Big to Small

The tendency to ask from big to small is the exact opposite of the technique of putting a foot on the door. The salesperson makes an unrealistic request from the other person. Naturally, this demand doesn't correspond to the real issue. However, the salesperson then makes a request that is smaller than the same person. People feel responsible for such approaches, and they will accept the offer. Since the request is small, by accepting it, people have the idea that they will help the salespeople and the technique of moving from big to small requests works.

6. Use of Reciprocity

Reciprocity is a term for the mutual progress of a business. When a person does you a kindness, you feel the need to do him a favor. This is one example of reciprocity. For example, if someone bought you a gift on your birthday, you would try to pay back that gesture. This is more of a psychological approach because people don't forget the person who does something for them and they try to respond accordingly.

For marketers, the situation is slightly different from human relations. Reciprocity takes place here in the form of a marketer offering you an interim extra discount" or "extra" promotion... You are very close to buying the product introduced by the marketer you think offers a special offer.

7. Making Limits for Interviews

Setting a limit for negotiations is to provide an approach that will affect future rights. This is particularly effective when negotiating prices. For example, if you are trying to negotiate a price to sell a service, it might make more sense to start by opening the price from a higher number. Opening from a low number is not the right method because you have weakened your stretching share.

Even if the limitation for negotiations is not always useful, it's particularly useful in terms of price negotiation. Say the first number and get on with the bargaining advantage.

8. Limitation Technique

Restriction technique is one of the most powerful methods to influence human psychology. You can see this mostly in places selling products. For example, if a store has a discount on a particular product, it may limit it to 500 products. This limitation can be a true limitation or a part of the limitation technique. So you think that you will not find the product at that price again and you agree to buy that product at the specified price.

The restriction technique is particularly useful in new products. As soon as a new product goes on sale, you can convince people to buy it for a limited time or by selling a limited quantity of products with extra promotions or discounts. People who think that the product will not be sold again at a similar price may choose to buy the product you have chosen thanks to the success of your persuasion technique.

Persuasion techniques are not limited to these. Different techniques can provide more successful results in various fields. However, most of the techniques we may encounter in our daily lives consist of the methods shown here. If you want to be a marketer, if you are trying to sell a product or service, you need to have detailed information about these techniques if you want to make them available.

Chapter 9

How to Use Dark Psychology in Real Life

People use psychology within their daily lives, so why not use Dark Psychology and the tactics to protect yourself in everyday life. There are quite a few personality traits that can be very harmful if you get caught up in them. Sadists fall under this category. For instance, this personality type enjoys inflicting suffering on others, especially those who are innocent. They will even do this at the risk of costing them something. Those who are diagnosed as sadists feel that cruelty is a type of pleasure, is exciting, and can even be sexually stimulating.

We do have to face the fact that we manipulate people and deceive people all the time. When it comes to deception, people are deceiving not only others on a daily basis, but they are also deceiving themselves. People often lie to gain something or to avoid something. They might not want to be punished for an action, or they might want to reach a goal, and they self-deceive to get there.

Here are some examples of how people can deceive themselves:

Having a hard time studying - this is a common occurrence. When people are trying to study, they find a lot of things that can distract them, especially cell phones and social media apps. They will find just about anything to distract them from the task at hand. These types of people seem to have a phobia of not studying long or well enough and they are afraid that they will come home with a bad grade and it will show how unintelligent they are. So, they take the art of self-deception and come up with the idea that will help prevent them from studying. This excuse will weigh better in their mind if they do end up getting a bad grade on their test. The person's subconscious is telling them that it is better for them to get bad grades for lack of studying than to study and failing and therefore having to blame their intelligence. They couldn't live with that.

Here are other ways that we regularly deceive ourselves:

- Procrastinating – People often waste time when they do not want to study or do something important. However, the main reason for procreating could be the phobia against failing and procrastinating was just an excuse. Self-confidence can be an issue as well.

- Drinking, doing drugs and carrying out bad habits -People often fall into bad habits, drink, or do drugs just to have something to blame if they fall again. This type of person will try to convince themselves that if they could stop doing drugs, they could be very successful. When they are the ones deceiving themselves and standing in their own way.

- People often hold back because life is unfair. They tell themselves that we all live in a big lie that most people believe in, but not them. It is easier to blame it on life being unfair, then hold ourselves accountable for not reaching our goals.

If you realize that you have been deceiving yourself, here is a couple of things that you can do to change that.

- Remember that you are smart and the fact that you have been able to deceive yourself reaffirms it. If you were not smart, there would have been no way that you would have been able to come up with some of those ideas.

- It is important to learn how to face your fears. If you are running from a certain trauma, or not wanting to take a test, you have to remind yourself that you are stronger than this and that you can beat it.

- Lastly, once you face your fears, your self-confidence and courage will grow.

Manipulation in our daily lives

Manipulation is an underhanded tactic that we are exposed to on a daily basis. Manipulators are people who want nothing more than to get their needs met, but they will use shady methods to do so.

Those who grew up being manipulated, or being around manipulation, find it hard to determine what is really going on because if you are experiencing it again, it might feel familiar. Maybe you were manipulated in a relationship, or the current relationship that you are in reminds you of your childhood.

This is important because manipulation tactics break apart communication and break a person's trust. People will often find ways to manipulate the situation and play games rather than speaking honestly about what is going on. However, others value communication only to manipulate the situation to reveal the weaknesses of the other person, so that they can be in control. These types of people do this often in conversation. They have no concern with listening to others talk about anything about themselves. And they are not there to help those people get through whatever it is that they are going through. It is all about dominance in this case and that's it.

Here are some of the tactics that can be used on an everyday basis:

. Some of the common techniques that we can experience are:

- Lying – White lies, untruths, partial or half-truths, exaggerations, and stretching the truth.

- Love Flooding – Through endless compliments, affection or through what is known as buttering someone up.

- Love Denial – telling someone that they do not love you and withhold your love or affection from them until you get what you want.

- Withdrawal – through avoiding the person altogether or giving them the silent treatment.

- Choice Restriction – Giving people options that distract them from the one decision that you don't want them to make.

- Reverse Psychology – Trying to get a person to do the exact opposite of what you want them to do in the attempt to motivate them to do the direct opposite, which is what you really wanted them to do in the first place.

- Being Condescendingly Sarcastic or Having a Patronizing Tone – To be fair, we are all guilty of doing this once in a while. But those who are manipulating us in conversation are doing this consistently. They are mocking you; their tone indicates that you are a child, and they belittle you with their words.

- Speaking in Universal Statement or Generalizations – The manipulator will take the statement and make it untrue by grossly making it bigger. Generalizations are afforded to those who a part of a group of things. A universal statement is more personal.

→ Example: Universal Example: You always say things like that.

→ Example: Generalization: Therapists always act like that.

- Luring and Then Playing Innocent – We, or someone we know, is good at pushing the buttons of our loved ones. However, when a manipulator tries to push the buttons of their spouse and then act like they have no idea what happened. They automatically get the reaction that they were after and this is when their partner needs to pay close attention to what they are doing. Those who are abusive will keep doing this again and again until their spouse will start wondering if they are crazy.

- Bullying - This is one of the easiest forms of manipulation to recognize. For example, your spouse asks you to clean the kitchen. You don't want to, but the look they are giving you indicates that you better clean it or else. You tell them sure, but they just used a form of violence to get you to do what they wanted. Later they could have told you that you could have said no, but you knew you couldn't. It is important to note that if you fear that you cannot say no if your relationship without fearing for your safety, then you need to leave the relationship.

- Using Your Heart Against You – Your spouse finds a stray kitten and wants to bring it home. The logical thing to do would be to discuss being able to house and afford the cat. But instead, they take the manipulative approach. Their ultimate goal is to make you feel bad about not being able to take care of the animal. Don't let anyone, even your spouse, make you feel that you cannot make the best choice for you. You do not have to take care of the kitten if you don't want to. Bottom line. Meet their manipulations with reasonable alternatives.

- "If you love me, you will do this" – this one is so hard because it challenges how you feel about your spouse. They are asking you to prove your love for them by giving them what they want from you, making you feel guilt and shame. The thing you can do in this instance is to stop it altogether. You can tell your spouse that you love them without having to go to the store. If they wanted, you to go they could just ask.

- Emotional Blackmail – this is ugly and dangerous. The idea that someone will harm themselves if you leave them is harmful at the core. They are using guilt, fear, and shame to keep having power over you. Remember that no one's total well-being is your responsibility alone. You have to tell yourself not to fall for it. This will always be a manipulation tactic. However, you can tell them that if they are feeling like they are going to harm themselves that you will call an ambulance to help them.

- Neediness When it's Convenient – Has your spouse started to feel sick or upset when they didn't get what they wanted? This is a direct form of manipulation. For instance, they don't want to go somewhere with you and have a panic attack, that you have to help them through, so that they don't have to go at all. This is not healthy at all, and if this persists you should think about ending the relationship.

- They Are Calm in Bad Situations – When someone gets hurt, or their conflict, somebody dies, your spouse always seems to not react with any feeling. They are always calm. This type of manipulation makes you think that perhaps how you are reacting is a bit much. Maybe your emotions are a little bit out of control. This is a controlling mechanism because no one should be able to tell you how to feel. This might seem like they are questioning your mental health and maturity level, and you find yourself looking to them and how to respond in certain situations. If this something that happens often and you see that you keep falling for it, you might need to go and see a therapist. This way, they can help you work on your emotional responses and find your true feelings again. This manipulation method can be very damaging to your psyche. At the moment, learn to trust your gut. It will not steer you wrong.

- Everything is a Joke – This is a two-part manipulation tactic. Your spouse will say hurtful things about you, and then when you get upset, they get upset because you can't take a joke. Other times they will joke about you in front of others, and if you don't respond positively, you are again ruining the fun. This is a way to put you down continuously without having to take responsibility for it. Remember that you are not ruining the fun here, but you have to stand up for yourself.

Chapter 10
Why Analyze People

Have you anytime looked at someone and thought you had them understood just from that look? Is it exact to state that you were right? Or then again would you say you were stirred up about some piece of their character? Despite whether you were right or wrong, you essentially tried getting someone, which is an ability that most of us would love to have. Everything considered, in case you can tell when your chief is feeling incredible, you understand when to demand a raise, right? When you understand your people are feeling awful you know, it is anything but a chance to unveil to them you scratched the vehicle. It is connected to appreciating what understanding people means and how its capacities.

What Is Reading People'?

When you look at someone and feel like you can condemn whether they are feeling extraordinary or a horrendous one, paying little heed to whether they are a wonderful individual or a mean one or whatever else using any and all means, you are getting them. At the point when all is said in done, understanding someone means researching them and it does not just should be a speedy look, and knowing something about them without them saying anything in any way shape or form. It is a tendency you get from looking and from viewing the way in which they stand, the way wherein they look around, the way where they move. There are some different features that could play into your inclination and cognizance of them, yet the most critical thing is that they did not explicitly uncover to you whatever that thing is.

By and by, various people investigate someone and acknowledge they know something. You mull over inside 'charitable, they look sincere' or 'astonishing, they look upset.' These are instinctual suppositions and thoughts that we have when we see a person. As we speak with them, we may achieve new goals or even as we watch them over the room. Maybe you never banter with that individual, anyway you have examinations and considerations in regard to the kind of person that they rely upon what you have seen of them. You are getting them, and whether you are right or wrong is an assistant point.

For What Reason is Reading People Important?

For what reason would it be a smart thought for you to essentially disturb getting people? Everything considered, there are a couple of special reasons this can be a better than average capacity. In any case, at a most fundamental level, it reveals to you how you should approach someone. If they look neighborly, you might also be prepared to approach with a smile and a very much arranged welcome. If they look down and out, you might undoubtedly approach with a reason rather than basically stopping to make appropriate associates. If a friend looks upset, you may ask them what's going on or what happened. Understanding what they feel like just from a quick look can empower you to imagine whatever is going on essentially like that, and the better you get with the mastery, the better you'll be at talking with people.

In case you do not have the foggiest thought about how to scrutinize people in any way shape or form, you could wrap up interpreting something that they do or an action or an outward appearance mistakenly, and you may start to expect things about a person that is not correct. Maybe you see their face and accept that they are a perturbed person when they're basically furious with a condition. Maybe you think they look threatening, anyway they're basically perplexed with something that is going on around them. By making sense of how to scrutinize better, you'll have the alternative to push your life from numerous perspectives.

Understanding people can empower you to acknowledge who to approach with that unprecedented new idea (and when to approach) and who you ought to stay away from. It is like manner discloses to you how to familiarize something with them, paying little mind to whether from a precise perspective or dynamically fun and creative one. Before you know it, understanding people will be normal to you if you practice it routinely enough. Additionally, what's shockingly better is that you have no doubt been doing it for as long as you can remember and not despite contemplating it. That is in light of the fact that it is something that even kids will give a shot every so often, without acknowledging how huge it is.

Understanding People in Childhood

When you were an adolescent did you ever sit on a seat at the entertainment focus or on your porch and watch people walk around? You apparently did at some point or another, paying little mind to whether it was uniquely for a few minutes. Also, a short time later you look at the overall public and make stories. On the occasion that they're walking a canine, perhaps they're a pooch walker on their way to the entertainment focus. On the occasion that they're passing on an organizer case and walking quickly, they are late to a noteworthy social affair, clearly, that get-together may have been with outcasts in your young character, anyway, you get a general idea. You have successfully deciphered what you see of someone to make a story about them.

As you get progressively prepared, you use those proportional sorts of aptitudes to start understanding people extensively more and to some degree all the more accurately. Your cognizance of outward appearances and position start to develop to some degree more and before you know it you can look at someone and rapidly acknowledge what it is, they are feeling at any rate, as a general rule. All things needed are a touch of producing for your childhood capacities and before you understand it you are en route to progressively significant accomplishment in your adult life.

Getting Help Reading People

Understanding people is a noteworthy ability to learn. For a large number of individuals, you probably look at is an 'acknowledge the main decision accessible' circumstance, is not that so? You accept on the off chance that I can scrutinize people, at that point extraordinary, yet if I cannot, well, no harm was done, is that not so? Everything considered, really understanding people energizes you a lot in your life and it causes you to be a prevalent individual as well, which is the reason it is a critical ability to have, paying little respect to whether you have a straightforward appreciation or an undeniably expansive one.

If you do not perceive how to examine people, it is a capacity that you irrefutably can learn. It is something that you can tackle for yourself by fundamentally coming back to those extensive adolescent stretches of making stories for the overall public walking around. And yet it is something that you can develop impressively further if you push yourself. The key is guaranteeing that you do not stop and do not desert the progression you are making. You may be bewildered precisely with the sum you can learn in a short proportion of time in case you move on these capacities, despite starting with people you certainly know.

For the people who are not sure where to start or how to wear down getting people, it is absolutely possible to get capable help with the strategy.

Starting with People You Know

It will, in general, be less complex to start scrutinizing the all-inclusive community you know before continuing forward to untouchables. These are people that you certainly know things about, and when you look at them, you can in all probability watch things that show those qualities. If your nearest friend is excessively bubbly and pleasant to everyone, you can undoubtedly look at them and jump on that trademark. Venture up to the plate and see them, see what it is about them that shows others they are bubbly and all around arranged and a while later quest for those characteristics in different people around you.

It moreover empowers us to move appropriately between our own one of a kind perspective and another. Unusually, social understanding relies upon information that cannot be truly observed at this point ought to be translated from moving toward information and our knowledge into the social world.

Moreover, continuously, confirmation proposes social cognizance incorporates reenactment, copying others' experiences as a way to deal with getting them. A real model here is the manner in which we experience other people's sentiments.

When watching someone's face we will, as a rule, duplicate her outward appearance, smiling when she does, glaring in comprehension. Such mimicry may not be obvious to the nice onlooker, yet minute muscle order can be distinguished in all regards not long in the wake of being exhibited to an energetic verbalization. Surely, even our eyes extend so as to the ones we are looking.

Chapter 11

Hypnosis

Why is hypnosis done?

Therapists say that hypnosis is an excellent way to cope with anxiety and stress. For instance, if someone is supposed to go for a medical procedure that they are anxious and stressed about, hypnosis can help calm them before the procedure. There are various conditions where hypnosis is used. These may include:

- Pain control – if a person is suffering from chronic pains from cancer, childbirth, joints, headaches among others, hypnosis may help in bearing the pain.

- Hot flashes – when a woman is going through menopause, she will experience hot flashes that are uncomfortable most of the time. Hypnosis has been known to help with the discomfort of hot flashes.

- Behavior change – some people may find themselves having behaviors that are undesirable. Such include bed-wetting, insomnia, eating disorders, among others. The use of hypnosis has been known to help in transforming these undesirable behaviors.

- Side effects of cancer treatment – during cancer treatment, patients go through chemotherapy and radiation treatment. These forms of treatment leave the patient with undesirable side effects. The use of hypnosis helps cancer patients deal with these effects and cope with the treatment.

- Mental health conditions – many people suffer from various mental health issues such as post-traumatic stress, anxiety, phobias, among many more. The uses of hypnosis help a person deal with these conditions and bring relief.

Preparing for Hypnosis

There is no need for special preparations before a hypnosis session. However, it is recommended for a person to be comfortable and relaxed. It is also important for a person to be well-rested to avoid falling asleep during therapy.

Before you go for therapy, research and ensure the therapist you have settled on is certified to perform hypnosis. Look for someone you trust that has undergone hypnosis and ask for references.

You can also opt to interview the therapist before the session by finding out some of these:

- Does your therapist have psychology, social work, medicine training?

- Is your therapist licensed and certified to perform hypnosis?

- Where did the therapist get his training from?

- How much training does your therapist have in hypnotherapy and where did they get the training from?

- Does your therapist belong to any professional organization and if so which ones?

- How long has the therapist been in practice?

- How much the cost per session and are their sessions covered by insurance?

Once you have settled on a therapist, he or she will explain the expectations and the process. The therapist will also review your treatment goals with your help to ensure they have it right. The therapist will then embark on talking in a gentle soothing voice as he describes pictures that create relaxation, well-being, and security.

When you are relaxed and in a state of reception, the therapist will begin to suggest ways you can achieve your goals. A therapist may also help you have a vivid mental picture of yourself accomplishing the goals. Once the session is over, you can either bring yourself out of it or the therapist will help get out of your relaxation state.

During hypnosis, one does not lose control of their behavior. A person is always aware and remembers all that happened during the hypnosis session.

Hypnosis is used to help in coping with pain, anxiety, and stress and is used in cognitive behavioral therapy to change the behaviors and thoughts that are undesirable. However, hypnosis is not recommended for every person. Some people find it hard to get hypnotized while for others it is easy, and they enjoy the benefits.

What are the risks of hypnosis?

When hypnosis is done by a trained therapist or a medical practitioner, it is considered a safe addition and alternative treatment. However, in people with serious mental health issues, hypnosis may not be the best method to use. There are various reactions to hypnosis. However, these reactions are rare, and they include:

- The person may feel dizzy after therapy
- Experience slight headaches
- After therapy, a person may feel drowsy
- A person can be distressed or anxious
- In rare cases, hypnosis can create false memories

Three Stages of Hypnosis

Hypnosis is a process that involves the deep body and mind relaxation. Before we get to the various hypnosis stages, it is important to first understand how hypnosis works or the process of hypnotherapy.

1. Getting ready – every hypnotherapy session with a qualified therapist must be carried out in a relaxed, safe and calm environment where there are no interruptions of any kind. There is usually a preliminary discussion between the therapist and the person to be hypnotized. This is usually done to establish if the person has had prior hypnotism sessions and their experiences as well as trying to establish the problem one needs working on.

Most of the problems usually include a behavior or thoughts a person needs to balance or completely changed. For instance, a person may be struggling with bed-wetting; this behavior with the help of hypnosis is addressed and changed.

A skilled therapist should gather as much information as possible during the preliminary talk. This is important so that he may work on the best technique for the particular person and problem. The pattern most therapists use during the session is loose. It follows:

- Preparing and screening a client

- Inducting a client to an altered state consciousness state

- Deepening the trance state that opens suggestibility

- Posthypnotic suggestions. This is where advice is given regarding the problem the therapist worked on.

2. Induction – in a typical hypnotherapy session, the initial 15 minutes are for helping the client relax their mind and body. This stage is referred to as the induction stage. It involves helping a person to enter into a light state of trance by the use of relaxation techniques that work on the mind and body.

Gradually, the person is encouraged to relax their muscles and mind. This technique is aimed at ridding a person of any tension and releasing anxiety. The therapist focuses on instructing the client to slow and control their breathing. This is also to help relax and distract the conscious mind so that a person focuses on the subconscious mind. Because of many methods of induction, it is important for the therapist to understand their client and apply a method that works for them.

3. Deepening a trance – this stage is where the subconscious mind is made ready to be more receptive to suggestions or new behavior. Once the mind accepts new thought patterns, a change in behavior follows. To deepen the trance, some therapists may opt to continue reinforcing the induction method used. The method can be accompanied by visualization techniques that are very deep to increase the trance. A qualified therapist knows that it is important

for a person to be deeply altered in consciousness before starting hypnotic suggestions.

Now that you know how hypnosis works, it is important to understand the three stages of hypnotism.

Stage 1 – Hypnoidal State

This is the stage of light induction. At this stage, the person is encouraged to relax and have an internal focus. This stage is light and is characterized by the fluttering of the eyes of the person.

Stage 2 – Cataleptic State

This is the stage where the therapist moves to deepen the trance state. To know if a person is in this state, their eyes move from one side to the other.

Stage 3 – Somnambulistic State

This is the deepest stage in a trance. This is evidenced by the rolling up and down of eyes. This is the stage where suggestions are given and received at a subconscious level and the person in some cases may not remember hearing them.

Chapter 12
Mind Control

The History of Mind Control and Its Effects Today

When reflecting on the history of mind control, you may think of brainwashing techniques used in prison camps and dangerous cults that have such a detrimental effect on people's minds to the point of permanent harm or death.

Headlines of mass suicide or long-term psychological impairment, Stockholm syndrome, or post-traumatic stress disorder may also come to mind. In everyday life, mind control is just as prevalent as always, though we may not always be aware of it or recognize the signs. The effects of mind control are not always obvious, and often, they influence our decisions, thoughts, and feelings in ways that we are not always aware of. Throughout history, mind control has been used as a means to instill fear and produce obedience among groups of people and can also be used within smaller groups or between individuals to yield powerful control over someone. When this happens, the power dynamic becomes severely imbalanced in favor of the manipulator(s). In countries or regions where people have very little freedom or liberty, certain regimes may have a stronghold over their citizens, by using the threat of imprisonment, punishment and other withholding fundamental rights as a result? Severe impoverishment and lack of proper food and water can often keep people in fear of disobedience or speaking out, for fear they may lose what little they have access to for their families and communities.

Today, mind control is widespread as it has always been. It occurs worldwide within governments, organizations, and between smaller groups and individuals.

In many ways, it's more obvious and present than ever, though we often ignore the signs. Commercial influence and the ability to convince people to buy products they don't need are powerful, especially when people are willing to go into debt or sacrifice their hard-earned income for something less important. Some forms of media and publications may often broadcast or publish certain headlines and events more often than others to provoke a sense of fear or urgency about home invasion or public safety. They may use emotive words and phrases to evoke responses of fear or shock, which causes people to live more cautiously and carefully, without deviating from the "norm."

What are the Signs of Mind Control?

Like manipulation, mind control aims to persuade and influence a person or people's ways of thinking, acting, and behaving to gain a benefit. When people are easily influenced and manipulated, they become more susceptible to practicing or doing things that would normally not consider as an option. The effect or success of mind control can vary depending on the techniques used, the target(s) and the environment. These factors, among others, play an important role in how successful and powerful mind control can be and also provide information on how to spot these signs before they develop further:

Isolation

This may seem like a severe case of solitary confinement, though isolation can refer to simply keeping you from friends and family.

This tactic is often used by an abusive partner or spouse to keep their partner away from the comfort and support of friends and family. Isolation can be psychological, in that the manipulator will gradually convince you that one or two family members are trying to control you when they are the one doing the controlling. Over time, if they are successful in persuading you that your family is deceptive or manipulative, they may continue to target friends and co-workers or acquaintances as well, telling you there is something wrong with them, or making you feel as though your friends are insincere, jealous or not truly worthy of your friendship. After a while, friendships and family members fade into the background, and you find yourself more emotionally dependent on the person practicing these mind control techniques. Isolation can effectively keep you from seeking help when you finally realize the dangers of being left alone with someone who does not have your best interest at heart.

Recognizing the early signs of someone or an organization to isolate you from others, even subtly, is vital to avoiding a long-term disaster. Any type of discredit or negativity towards good friends and family should be regarded as a possible sign of control. This tactic will usually occur early in a relationship, where the manipulator realizes a strong bond between you and others.

They see this as a threat to their ability to control you and will do anything in their power to break these relationships to keep you vulnerable to their will. If a group or organization appear inclusive and friendly yet questions the nature of your personal relationships and friendships, it's a sure sign they are seeking to gain more control of your life.

Mood Swings and Erratic Behavior

If your partner becomes easily agitated or angry when you disagree with them or makes you feel unworthy of their affection for expressing an honest opinion, they are grooming you to bend to their will. For the manipulator, there is little or no room for any variance in opinion or thought. They will only accept complete submission and agreement.

Anything less will result in erratic mood swings and unpredictable behaviors. In extreme cases, some manipulative people become violent or aggressive. The very threat of this possibility will convince their victims to remain obedient simply out of fear. Recognizing severe changes in mood or emotion, especially when there is no reason or event to trigger the change, is a good reason to avoid someone. Over time, this behavior will escalate and become worse, especially once you discover their tactic and need to escape their manipulate grasp.

No Compromises

Mind control, when effective, requires complete and total obedience. There is no room for other thoughts or compromising. In a healthy relationship, all opinions expressed are regarded with respect, even if disagreements or debates are surrounding certain topics. Not allowing another person to express their thoughts without ridicule or judgment can convince them that they are not worthy or compromise. It is also a form of psychological and emotional abuse. This is an easier sign to recognize when even the smallest of decisions or ideas are bent to the will of the manipulator. This can mean anything from choosing a restaurant for dinner or film to watch, which later affects more significant decisions, such as a mortgage or starting a family. Knowing when to spot a lack of compromise can save you a lot of grief later in life, especially where a long-term relationship may form.

Who Uses Mind Control? Organizations, High Control Groups, and People We Know and Encounter in Everyday Life

Who uses mind control, and for what purpose? Many people who are susceptible to the influence of mind control don't often realize they are. There are different people (individuals) and groups that employ mind and thought-control techniques for a variety of reasons.

Understanding their purpose also provides a good explanation for why certain techniques are used and how to recognize them. When we encounter everyday situations, from commuting to work or school, shopping in a grocery store or running errands, we may experience a form of influence or covert manipulation through a sales pitch or billboard ad, without realizing its effect.

If we stopped every time, we noticed an ad, a promotion, a person or representative attempting to "pitch" a sale or ask for a donation, we would then realize how bombarded our mind is with persuasion. In reality, only certain ads or people will catch our attention, while others will slip away.

Chapter 13
Body Language

Body Language

Language is a blunt force tool. What is language supposed to do? It is supposed to convey thoughts, ideas, concepts, and stories to other people accurately. It gives us a way to interact and puts us all on one level of communication so that we can exchange simple messages to each other and get by. However, language is also responsible for transmitting the most important, deep, and abstract concepts. What it comes down to is the complexity of our everyday experience. How would you describe the flow when you're replaying basketball and making every shot? How would you describe that in words to someone and have them know what you were experiencing? What about when you eat a piece of chocolate? Get broken up with? These are things that can't be described in words, and yet we try to describe them. Sometimes it is done in ordinary conversation; sometimes, it is done in art or literature.

Language is what mutes and bottlenecks our experience into what we can convey to other people. Language is so limited in its ability to truly share our knowledge with others, and it is that limitedness that makes it, so that body language is so important. You are often experiencing both at the same time; you are experiencing someone's language simultaneously with their body language. Non-verbal communication is not all just body language, but a huge part of it is body language.

Think about one interaction that you have had in the last few days. It could be anything from buying something at the store, to a wedding—any tiny little interaction that you had. Try to imagine the interaction from the very beginning.

Body language is comprised mostly of a few factors: effect, posture, and motion. The effect refers to a person's facial expression. If a person is smiling, you could say that they have a bright effect. A person's effect is not always congruent with what they're saying and experiencing. You might see this in someone who is talking nervously about something, and they begin to smile. This means that their expression does not fit whatever they are talking about and that there is incongruence in their effect. When a person has a congruent effect, their facial expressions will change and be malleable. A person who has a congruent and secure effect will be expressing whatever they're thinking about or talking about on their face.

Posture is the way that a person holds himself or herself. This comes from their orientation to the world. This can be found in the Enneagram of personality that we talked about earlier. These personality types describe an orientation to the world. Some people are rear oriented as warriors; others are oriented as perfectionists. The way that a person's personality is will dictate the way they hold themselves physically. A person who is up in their mind will have the posture of a distracted person. A proud person will lead with their chest. What the chest symbolizes is a place of pride. It is where the heart and lungs are, and a cage of bones protects ties it's a very important part of our bodies, and when we lead with that, we are showing that we are confident.

What would the chest be doing on a person who did not feel confident? If a person is not confident, they will not walk with their chest leading, rather it will be in a collapsing position. Think about a person who is not confident, and how their shoulders move forward, and their posture seems tired or broken. They are the ones without confined because they are trying to protect their heart.

Another aspect of nonverbal communication is art. When we talk about art in this sense, we are talking about the capital "A" Art that includes sculpture, writing, acting, and all creative arts. Even when language is involved, it is not in verbal communication; it is in writing. All of these fall under nonverbal communication. Learning to participate in artistic creation can help you to be a person who is more in touch with this part of communication.

Art can have all kinds of functions. Sometimes its function is to help sell things. This is a form of communication. When you hear something on the radio that is a catchy jingle that makes you feel a certain way about a product that is a deep form of communication. Art can help us to dance, think, feel joy or sadness, help make things clearer, make political action, call to war, call for peace, etc. Music has a variety of functions. It can be used to help us energize or relax. Art is the same way.

Non-verbal communication happens all the time; you just don't notice it. Your gaze has a deep implication on how people perceive you, the way you walk can tell people a whole lot. A person is a private being when they do not show you much with these.

Some people do not have the power of verbal communication. Some people have very advanced dementia or other mental disorders such as advanced schizophrenia. Others could be people with learning or developmental disabilities like autism or Down syndrome. Can these people still communicate? Absolutely! They can communicate because their lives have revolved around learning ways for them to communicate.

Some ways through which one can easily communicate with people who are non-verbal are touch, music, art, or hand symbols. People who are non-verbal tend to experience deeply.

Some people have learning or developmental disabilities that prevent them from reading non-verbal cues. People with Autism Spectrum Disorder have a hard time deciphering the cues of behavior and non-verbal communication. ASD is a somewhat mysterious condition, and it is only diagnosed and marked by certain behavioral patterns and lack of social ability. This makes it a fascinating condition to learn how to help people with ASD to function better. For kids with ASD to be able to function better, they have to be assisted with integration. This means they must learn to use their sensory inputs in concordance with their cognitive abilities to learn what a person is expressing. They will have to learn that when a person has their face all scrunched up, and they are yelping, that a person is angry. They have to learn about the body language of a sad person and how to act around that person.

Rather than teaching kids how to learn the basic cues of non-verbal communication, we are trying to encourage you to learn to trust your intuition and be able to analyze behavior patterns on a deeper level.

This means that when you experience a behavior pattern, you are able to surmise what this means for you and what it means for other people around you. Instead of thinking about your feeling and worrying about it, you can either express it or act on it or do whatever wiles they need to do.

This is where the intuition comes in. You've got to trust what you are feeling about a person. If you see that a person walks into the room with a smile you've known before, and they act a certain way that you saw a person act, and you can know that they are trying to deceive you, this will make your life a little bit easier, as you will have that knowledge going in.

If you just started attending a church, and at first you like it because of the community, but then you start to feel that it is just not the right place for you, this is intuition. We can use intuition to the behavior patterns of others to know if they will be good partners, good friends, etc.

Let's use the example of dating to try and illustrate what we are talking about when we talk about intuition. A new partner will be a new experience. It will be something that comes to you when you need it.

People that we get involved with are generally on the same level of personal development as we are. If they aren't, these will inevitably lead to tension in the relationship. When we get into a romantic relationship with a person, we start to blend our patterns of behavior. This means that you will seek out a person that will tend to increase the behaviors that you want to improve within yourself.

This is a good and bad thing. It is a natural process that lets us select people to get into relationships with so that we function better in our lives. However, if we are not able to see how we aren't functioning well in our lives, then we will just be looking for someone to help us continue the patterns that we already find so easy to do. This is how patterns in relationships are perpetuated.

Various aspects of physical expressions

- Eyes

Those who have studied NLP would agree to the adage that the eyes are the windows to the soul. You can read another person's true emotions by observing how their eyes move or how fast they blink. Even though the eyes are relatively small as compared to the other body parts, the eyes can create numerous expressions that can reveal the entire truth.

Let us start by telling whether or not a person's way of blinking is normal. You have to assess whether they are blinking too fast or too slow. If someone is anxious about something, they tend to blink fast. Moreover, rapid blinking also indicates that they are telling a lie. On another note, if the person is blinking slowly, it shows that they are trying to control their eye movement. You can assume that they are hiding something or trying to suppress an emotion.

How the eyes move also shows how engaged the other person is in the conversation. Those who show interest in the conversation would rarely break eye contact. If an individual frequently breaks eye contact, they may be distracted or bored. Moreover, it can also mean that they are naturally submissive or may be nervous about conversing with other people. Whenever you are talking to someone whom you are trying to decipher, pay attention to how they glance at certain objects in the room. Anyone who barely maintains eye contact and always glances at their watch or at the door is secretly saying that they wish to end the conversation and leave.

The size of the pupils can also reveal how much a person is interested in the conversation. Now, determining whether or not the other person has dilated pupils can be a bit of a challenge even under the right conditions. Moreover, you have to consider that a dilated pupil may have been affected by light. These are the things you have to consider when trying to decide if a person is genuinely interested. The good news is that you are always allowed to test this out. Remember, the dilation or contraction of one's pupil is automatic, so you have to have a keen eye. Next time you are conversing with someone who appears to be bored, immediately switch the topic into something you know elicits interest from the other person and observe the change in their pupils.

- Voice

If we are talking about body language, then it is crucial that we discuss the use of vocalic. It is important to note that vocalic is different from the words that pass a message.

Chapter 14

How To Recognize A Harmful Relationship

In a romantic relationship, it is more than likely most common to identify an emotionally/psychologically abusive situation during the incident and reconciliation situation. An article written about domestic violence/abuse from Help Guide identifies questions one can ask about themselves or their partner:

- Do you feel afraid of your partner?

- Do you avoid topics out of fear of angering them?

- Do you believe you deserve to be hurt or not treated well?

- Do you wonder about your own sanity?

- Do you feel numb or helpfulness?

- Does your partner humiliate you often?

- Does your partner blame you for their abusive behavior?

- Does your partner put you down for your accomplishments?

- Does your partner see you as an object rather than a person?

- Does your partner force you into sexual intercourse?
- Does your partner threaten you with violence?
- Does your partner threaten suicide?
- Does your partner keep you from seeing friends or family?
- Does your partner constantly check up on you?
- Does your partner have an unpredictable temper?

If the answer to most of these questions is yes, then it is likely that you are in an emotionally abusive relationship. Realizing this is very important. The very first step in recognizing that the behavior your partner is committing is abuse.

The next step is realizing that it is not your fault. There is never any circumstance in an abusive situation where the fault lies in the abused. This is a fallacy that the abuser attempts to convince their victim over a long period of cyclic behaviors. This is a huge step and usually, don't have a fair amount of effort through the support of loved ones. This must occur by no longer making any excuses for the abuse; no matter how unintelligent, unreliable, or undesirable you believe you are, there is no excuse for behavior that is abusive. The abuser wants you to think that no one else will be able to love you as they do, therefore, you do not think leaving is possible or practical. This is the key to the narcissist's malevolent tactics.

Once it is recognized that you are being abused, and you are convinced that it is not your fault, it is time to begin documenting everything. Now, the concept of leaving is realistic and possible. This is very important if you have children or are married. Journal entries are a tactic, or even audio/visual recordings, if possible, are good ideas should it be available. If the abuse is physical, attaining a restraining order would be very important too. Keep the documents in a safe place that your partner cannot locate or even at someone's house that you trust.

Packing an emergency bag is very important. This too can be kept at a person's house in whom you trust. Keep anything essential to a couple of days or even a week of functioning, such as medication, identification, and money. Abusive situations can escalate very quickly, and you have to be ready to leave at a moment's notice.

Alerting your family and friends is a good decision once you or someone who is being abused can admit that they are in an abusive situation. The support of people you love and who love you is highly important during this difficult time. The abuser may have been masterful at separating you from those you love to avoid support of leaving. Once you realize you are being abused, trying to reach out to family and friends again is something that may be difficult, especially if the abuser was successful in keeping you separated. But stay dedicated and realize that the separation created is another tactic your abuser applies.

Finally, the act of leaving and completely disengaging from this partner is the most crucial activity to participate in to begin the process of healing. The act of leaving is both physical and mental. An abuser can react to this separation with unkind acts or flowers in an attempt to continue the cycle of abuse. You should realize that these behaviors are them trying to bring you back and keep you controlled. Make an effort to block the abuser on social media, email, and phone number, along with your restraining order. Remind yourself that this person injured you, whether it be physically or emotionally. This person is not going to stop their behavior. You cannot love them out of their abuse. The only way it will stop is for you to leave the situation and begin identifying your tendencies toward codependent behavior.

How to Avoid Manipulation

Beyond the realm of emotional abuse, there are various ways that you can detect how others are trying to manipulate you and how to avoid it. An article from the website The Power of Positivity lists 11 ways that this can be done.

1. Don't fall into their trap. People who try to manipulate others will try to do so in any way possible. If you cannot avoid these people, especially those who are in your family or who you work with, try ignoring them or responding with something kind, rather than falling into their trap of reacting to their prodding. Their goal is often to get a rise out of you, so if you don't react in the way that they want you to. They will eventually leave you alone.

2. Start writing down what they say during conversations. Emotional manipulators will often use what has been said or contort what

they believe has been said to benefit them. If you don't have this recorded, you could begin doubting yourself and believe the lies that they are telling you. If you start writing things down, you will have tangible evidence that they are trying to manipulate you.

3. They will more than likely become angry if you confront them with this information, but if you remain persistent and protect yourself from their reactions, they will stop trying to toy with you.

4. Steer clear. If it is possible, try to stay away from people that you feel are emotionally manipulative. If you are a person that feels you can read the energy of others, if you feel something negative coming off of them, try to stay away. Trust your intuition in knowing that this person is not good for your mental health. Do what you can in a workplace or familiar situation to avoid contact.

5. Call them out on their behavior. Emotional manipulators have more than likely rarely been called out for their abusive actions. If push comes to shove, you must stand up for yourself and inform them about how uncomfortable they are making you, and that you are aware you are being taken advantage of. Even if their reaction is greatly aggressive, at least you know that you indeed stood up for yourself and they know you are not going to take this kind of treatment anymore.

6. Avoid emotional attachment. This is easier to say when you first meet a person, and their true manipulative self-reveals itself early on. But if the person has successfully conned you, which is often the case in emotionally abusive situations, try your best to recognize that their behavior is not good for you, back away, and create healthy boundaries. The more attached you are to this person, the easier it will be for this person to manipulate you.

7. Meditate. Meditation can help you self-reflect and look at yourself and others with compassion. It is helpful while dealing with an emotionally manipulative person because nothing they do will affect how you are feeling. Again, this person may tire of you and move on. It is not your job to change the person, so don't worry too much about the story behind why this person is the way that they are.

8. Inspire them. If you are feeling so inclined, perhaps you can suggest some methods you used to help yourself becoming your own best self. Again, it is not your job to change this person, but making positive suggestions could turn around the energy of negativity they are so used to thrive on.

9. Tell them "You're right". This may be hard for your ego, but if you let the manipulator know that they are right, you are instantly cutting off the drama portion of what they feed off of. There is no space for arguments, so they will tire of you.

10.   Let go of harmful relationships. This is in relation to a romantic or close relationship with a person whom is very manipulative. Your mental health is important, and if you have tried to help this person and they continue abusing you, their fate is in their own hands. You should let go of this person and choose to find your way through life without being constantly emotionally drained and pained.

11.   Develop a strong mentality. Insults or ways that manipulators try to wind people up will not affect you if you choose not to let it. If you are insecure or not confident, try to admit this to yourself so you can move forward and develop this to avoid allowing anyone else to use or hurt you.

12.   Give yourself positive self-talk: A good mood can be completely ruined by a talented manipulator, so if they are making you feel particularly agitated one day, try various methods of self-talk with yourself. Try to remember that you are your person, and that your interpretation of yourself cannot be affected by a person who only has a selfish agenda. You can find various affirmations online or on apps on your phone. Like most of these tactics, you will eventually at least appear unaffected to the manipulator, so they will move forward and away from bothering you.

Chapter 15

Brainwashing

If you would ask a person at random whether they know what brainwash means, you get an affirmative answer 90 percent of the time. That's not the case more often. This concept has been heard by many of us, but we often tend to confuse our vague realization of what it is for an otherwise accurate understanding. The question we dare ask ourselves then is what brainwashing means and why it does. It is perhaps the essential kind of dark psychology we are going to discuss, which offers much more negative results than the rest.

This complex brainwashing enigma is essentially the gradual process of replacing the ideas of a victim with a manipulator about their being and of replacing them with new ideas that should adapt to one's own needs, and that can either narrowly take place. For example, a manipulator can control a person or adopt the same principle but a broad group. The circumstance of brainwashing is like the one where people believe they can be a call for a higher purpose to join a terrorist organization. In the majority of other scenarios, indoctrination appears to be effective. Most believe that brainwashing is the Hollywood fiction of someone who continually imposes certain concepts on their victims, and within a short period, the person becomes the manipulator for every bidding. More realistically, it is a process that generally changes an individual's perception of reality far from what they have had a very gradual but voluntary process. The fundamental trick is to make the victim feel as though they are always in control. There are several situations that make brainwashing possible, and often they are motivated differently by it. Let us discover these specific scenarios, starting with the situation of a cult. A cultic can be said to be an organization of people who often believe that one individual is considered as their leader in something larger than themselves. The leader is often characterized by his exceptionally great influences and a very charismatic behavior on his followers. The question that we now try to ask ourselves is, why is brainwashing a culture in those contexts?

The very basic appeal that these cults possess is that actual reality is only graspable if they decide to cultivate and follow the teachings in the factions. This is often something that people want to listen to because the truth is that today's world is a complicated journey often appears confusing, which gives some relief from a promise of fraternity and an unparalleled opposite. Brainwash resonates here with the idea of a new normal. It is because of this idea that cults used by different sect leaders to brainwash the members to accept their mostly strange doctrines at the same time, making them a dominant theme in their lives. It sounded incredibly strange to idolize a mere mortal to the extent that he actually adored it. But that's so' normal' in a cult that it doesn't seem to be a little bizarre for other members. Probably the most potent way to ideological brainwashing is through this process of social strengthening. Cults have the same DNA as drug distributors. Why is that? Why? Usually, a person finds a particular sect to find something he or she lacks in life. In the drug world, the same thing happens because a person often tries not to experiment freely. Cults do not make brain ash people members, as people have sought some form of fulfillment of a particular desire. It is this innocent desire to search and to be prepared, making them easy goals for brainwashing.

The context of ideologies would be another similar one, which must brainwash on its foreheads like cults. Their principal difference is that not so much the individual is the focus of ideology, but the whole idea. In principle, ideological brainwashing is stirred up by persons who have put absolute and full confidence in a plan. It is regarded as a very dangerous tactic because it goes beyond one person. Use an example to understand this better. When you look at the religious extremist groups, they worship their leaders like tomorrow is no one. Would it be enough to say that your' god' could be killed? This is a definite no because, as long as they are 6 feet below the surface of the ground, they will be praised and recognized as martyrs who died because of their ideologies.

This theme not only resonates in the cultural playground, but it also places me on a more innocent level. The brainwashing effect on fans is too often experienced by musicians. If you hear music often, a diehard fan of a particular musician will mark a reasonably young and impressive person with his sense of identity and happiness for a specific musician. You are frequently able to defend your famous superstar as if you knew the person personally. Some fans will even go to a higher level to harm themselves if their idols also imply that it is a cool thing to do. Can you imagine the effects this will have if used in a cult context if ideological washing on the innocent level is so dangerous? In addition to the general brainwashing scenarios, it can easily take form in a personal setting.

When you dive into the bowl of personal brainwashing, we find that the brainwashing process, as it is at the border, is also the slow and gradual replacement of the existing beliefs of a man with those placed by a manipulator to serve his benefit more effectively. The main difference between the two is that unlike a group to implement the new typical scenario, the manipulators, in this case, aim at establishing a profoundly personal relationship with the victim. This is considered to be stronger than the brainwashing group because the brainwashing tape can be continuously altered to match the psychological structure of the victim.

After you have had an insight into the real occurrence of brainwashing, how do the manipulators use this kind of dark psychology? Hollywood's one thing is to lay the groundwork for indoctrination. Each brainwashing movie typically begins with a focus on the mental state of the individual and the social circumstances surrounding him — this way of controlling the mind. Brainwashing is not a technique that every single person can smoothly perform. These manipulators are often the preferred victim who seems to be attempting to fill an absolute emptiness, particularly those who are turned on by certain life events by their current reality. There is no question of losing a loved one as the best example of this to be illustrated. The emptiness is often conveyed to people wherever they go. This person, in search of their being full again, will choose to flee and join religious extremist groups and become suicide bombers in pursuit of being entire. The brainwasher gives us assurance as a killing ideology.

Once a handler intensifies his aim from the Internet or in an individual, the brainwashing method has been launched. Contrary to the image, the actual representation of these unhappy personalities, which looks strange and culturally withdrawn from the fellow with a particularly distorted perversion as an intellectual washer, is north of this because they are only ordinary individuals with whom we communicate on a regular occasion, with a calm and pleasant feel. Slowly but certainly, the manipulator will operate challenging to create an atmosphere in which the person thinks he can trust him. It is accomplished by generating clever peripheral associations such as a shared interest in a specific sport and a similar musical relationship. After that, the manipulator will then move forward to create even more confidence. He does so by creating some grim experiences from the background, which are also associated with painful survivors. For instance, if the victims share in the loss of a wife, the manipulator will give them a history in which he, too, has lost a friend.

The next phase is recognized as a demonstration of utopias. What does that imply? What does that imply? The manipulative person regularly provides alternatives to any of the issues that the people face. It is a brainwashing method. This is generally performed informally, at first, to prevent any negative pressure interactions of the person. The perfect introduction is never any specific item, and it is either character; the manipulator often tries to transform his target as well as religion or a particular religion. It could be either terror or his stupid willingness to honor and validate. If procedures are started correctly, a specific person often seeks to gain perspectives into a greater knowledge of the alternatives. In certain situations, the manipulator will initially retain this data, emotionally forcing the person to operate to prevent it. This outcome is what occurs effectively.

Once the person who was subsequently introduced into this new belief system and appeared to be reacting in such a manner that the manipulator would like to, his real motives will be revealed very carefully. This concept is called "gradual disclosure." In essence, this is a method where ideas are presented that are often simple to comprehend before anything disputed becomes apparent. When we stare at religious terrorism, this is made completely clear. Recruits are frequently taken in by persuasive phrases like God enjoys them. After you're a wholly committed participant, the story requires a severe turnaround. What those rulers advise the participants is that the person is now crossing a line of no exchange, and nothing can be said to modify their minds because of this love.

I can bet that the question you are asking yourself right now is, why does the person consider it still okay to still communicate with his oppressor after the aims of manipulation have been disclosed? Usually, this is because of several factors. First of all, the participants experience a powerful feeling of pleasure and acceptance at this stage. This is because they took the time to build this solid base of friendship and helped them out of a challenging period in their lives. Secondly, the victim has spent a lot of time and money on the process so far and isn't right to let all of that drain back from everything. This is also referred to as the price mistake. Finally, because the manipulator has now collected a box of conflict of the perpetrators of most mysteries, the person decides to remain. They are cautious that this could be used at some stage at the moment against them. The problem of blackmail is not generally grasped to explain, particularly if they are endangered, why a person would be associated with them. Why is that? Why? The manipulator is willing to hide the truth often in a non-threatening manner as a master of deception. The manipulator will tell, "if I can no longer assist you, you get the greatest person, like a near family-member bug, you can go," because the manipulator already has a profound feeling of relationship and confidence in the subject, the blackmail image and power is usually seen to emanate from it.

Chapter 16
The Role of Defense

To avoid falling victim to manipulators, you have to build your defenses so that you are prepared for any manipulative strategies that they may try to use on you. The best way to build your defenses is by taking steps to improve your self-esteem and your willpower. However, as a point of caution, you should be very careful about how you build your defenses because you don't want to create restrictions that will keep you from living a fulfilled life.

For example, as you try to guard against manipulation, you can't act out of fear. You can't hide from the world just to avoid scenarios where someone might want to take advantage of you. Remember that the world is full of people with dark personality traits who may harbor malicious intentions, so acting out of fear won't protect you from anyone. In fact, it will just make you more of a target. As you build your defenses, make sure that you start on the premise that you are willing to confront manipulators head on, and you will never run away or recoil. If you act out of fear, you lose by default.

The steps to raise self-esteem: To help you build your defenses, we will discuss the eight steps that you have to take in order to raise your self-esteem and to increase your willpower by extension.

Acceptance

Acceptance is about assenting to the reality of a given situation. It's about recognizing that a certain condition or process is what it is, even if it's characterized by high levels of discomfort and negativity. It's about consciously submitting to the fact that something cannot be changed, and that its reality is not subject to interpretation. It's about making peace with the situation that you are in.

Acceptance is the opposite of denial. Even the most rational among us tend to be in denial about lots of things in their lives, which are settled facts in real sense. Denial can be a coping mechanism, one that can keep us from being overwhelmed by the reality of a given situation. However, denial does us more harm than good, because unless we can accept something, we can't change it, and we will be stuck looking for alternative interpretations and explanations for our prevailing circumstances.

Without acceptance, the door remains wide open for malicious people to exploit us. Take the example of a patient who is told that he/she is terminally ill. After seeking the opinions of several medical professionals and getting the same diagnosis, the patient is still left with the choice of either accepting or denying the situation. The one who accepts it will make peace, and try to make the best out of what little time he has. The one who stays in denial will become susceptible to tricksters who may offer "alternative cures," and he may end up losing all his savings paying such people so that in the end, he leaves his family with nothing. That is an extreme example, but it perfectly illustrates why acceptance is important in avoiding manipulation, even if the reality may seem too painful to accept.

The most crucial form of acceptance is self-acceptance. It refers to the state of being satisfied with yourself, the way you currently are. Self-acceptance is a kind of covenant that you make with yourself, to validate, support, and appreciate who you are instead of constantly criticizing yourself and wishing you were someone else. Most people have trouble accepting themselves as they are. We are all in a constant strive for self-improvement. We want to be more successful, to be wealthier, to be more attractive, or to be perceived more positively by others. Even the most accomplished among us have issues with self-acceptance.

In many ways, the desire to be a better version of yourself can be seen as a positive thing; it can help you study harder in school, work harder to earn a promotion at work or exercise more to get in shape. However, the problem is there is always room for improvement, so no matter how high you ascend, the dissatisfaction will always be there, and it will make you vulnerable to manipulation by people who want to take advantage of your desires.

To defend against manipulation, you have to accept your reality, and you have to accept yourself. People tend to think that if they accept themselves, they won't try to improve – that couldn't be further from the truth. Accepting yourself means owning up to your flaws, and that gives you control over your life. With self-acceptance, attempts at self-improvement would come from within, so when you decide to change, you will be doing it for yourself and not for anyone else.

Increase awareness

Increasing your awareness means having a higher level of alertness when it comes to understanding what's going on in your environment. It means paying close attention to your surroundings, and to the way, people behave around you. The higher your level of awareness, the better you will be when it comes to adapting to your surroundings and understanding the motivations of the people you interact with.

When you become more aware, you will be able to catch on quickly when people try to manipulate you. Many of us tend to be preoccupied with our own thoughts that we hardly ever notice the cues of the people we interact with. We tend to live life on autopilot, so when other people try to seize control over our lives, we only notice it when it's too late. If you increase your awareness, you will be equipped with the skills necessary to identify all the red flags, and you will be able to stop most manipulators on their tracks before they can do any real harm.

The first step towards increasing your awareness is to learn about the tendencies of manipulative people. Reading this book puts you ahead of the curve; you now know enough to be able to spot people with ill motives, but you should understand that the worst kinds of manipulators are very good at concealing their motives, so you have to keep working on increasing your awareness.

To be truly aware of manipulative people, you have to approach all interactions with some levels of skepticism. We are not telling you to turn into a paranoid person who doesn't let anyone in; we are just saying that you should take a deeper look at each person you interact with. Try to study their body language and their words, and try to see if they are trying to hide something.

Apart from increasing your awareness, you have to increase your self-awareness as well. Many people confuse those two things, but they are entirely different concepts. Self-awareness is about understanding yourself. It's about having a clear concept of your own personality. You have to examine yourself and figure out what your strengths and weaknesses are, what your values and motivations are, and what kind of thoughts and emotions you are likely to have in specific situations. Self-awareness helps you understand both who you are and how other people perceive you.

Self-awareness works as a defense against manipulation because when you know who you truly are, it becomes more difficult for someone to alter your thoughts and perceptions. If you have strong and well-articulated values, it becomes harder for a manipulator to get you to abandon those values. People who like self-awareness are more likely to be gaslighted or to be subjected to other forms of mind control.

If you end up in a relationship with a manipulative person, self-awareness can help you keep your identity. Manipulators will try to tell you what to think and how to behave, but if you are self-aware, you will experience cognitive dissonance, and your brain will push back against any attempts at manipulation.

Detach with love

Detaching with love is a defense against manipulation that is most commonly used by people who have loved ones who suffer from substance abuse problems. Even though it was conceptualized to help people deal with addicts, it can also work when you are dealing with manipulators.

Detaching with love is about showing love and compassion for others without taking responsibility for their actions. For example, if you have a family member who is a drug addict, the way it works is that you try to support them and encourage them to get clean, but you let them make their own decisions, and you let them suffer the consequences of their actions. If the addict doesn't come home, you don't waste your time looking for them in the seedy parts of the city, you stay at home, and you do the things that benefit you and make you happy.

The point of detaching with love is to stop trying to control other people's lives, even if you are doing it for their own good. The idea is that you accept that people are different from you and that they have their own free will.

Detaching from love can defend you from manipulation in many ways. There are manipulators who want to exploit you by making you responsible for them. We have mentioned several times that some malicious people will take the submissive position in a relationship because they want your world to revolve around them. They want you to give them all your attention; that is how they control you.

When you detach with love, you will learn to stop fixing everyone's problems. So, when the manipulator tries to play the victim in order to gain your sympathy, you will keep doing whatever is in your best interest, and you will tell him or her to take responsibility for his or her own actions.

Some manipulators may take up self-destructive habits because they want to dominate you by making you clean up after them. When they do this, you can detach with love by letting them follow the paths they have taken, no matter where they lead them. If they are causing you harm, you can get away from them, but leave your door open. If they find the right path in the future and regain control over their own lives, you can let them in again. You have to make it very clear, through your words and actions that you will let them direct their own lives, and you won't take any responsibility for them.

Detaching with love is about accepting others for who they are, and respecting them enough to let them be in charge of changing their own lives. When you feel responsible for someone, and he makes a choice that harms you both, oftentimes, you will react with fear, anger, or anxiety. To detach with love, you have to learn to let go of those negative emotions.

Manipulators count on the fact that you will react in a predictable way to their machinations, but when you detach with love, you learn to calm yourself down and think about your role in the other person's life before you take any sort of action. This will keep you from falling into the traps that manipulators will set for you.

Detaching with love builds your self-esteem because it allows you to put your own needs ahead of those of the people that try to manipulate you.

Build self-esteem

You can defend against manipulation by building your self-esteem in the old school way; using self-help techniques. People tend to discount classic self-help techniques, but they actually work. They won't solve all your problems, but they'll make you feel worthy enough and give you the strength to resist many forms of manipulation.

Chapter 17

What Is NLP (Neuro Linguistic Programming)? A method to enhance personal development

The acronym NLP stands for neuro-linguistic programming, and indicates a methodology for changing the thoughts and behaviors of one or more people, in order to help them achieve the desired results.

Born in the 70s in California in the middle of the New Age era, NLP owes its success to the promise (often kept) to improve performance at work and to achieve happiness through personal development. The founders of NLP, the psychologist Richard Bandler and the linguist John Grinder, started from the belief that they could identify the thought patterns and behaviors of successful individuals, and then teach them to others.

One of the main techniques of NLP is in fact constituted by imitation or, as adepts define it, modeling: by imitating the language and behaviors of successful people it would be possible to make our skills our own and achieve their own results.

NLP is mainly based on language processing, but also uses other communication techniques to make people change their thoughts and behaviors.

How does it work?

NLP is based on the idea that people operate through internal "maps" with which to represent the world.

NLP thus tries to identify these maps (which are nothing more than subjective experiences of what surrounds us) to change their orientations. It is a methodology that aims at a change of thought and behavior.

It should be specified that NLP has nothing to do with hypnosis. On the contrary, it works through the conscious use of language to modify a person's mental and behavioral pattern.

What is it for?

NLP finds a wide field of intervention, using various techniques according to the desired purposes.

Starting from the idea that thought and behavior can be modelled, NLP is used for:

- treat anxiety, phobias and stress, thus improving emotional responses to certain situations;
- achieve successful professional goals, such as increased productivity at work and motivation;
- remove negative thoughts and feelings associated with a past event;
- improve their communication skills.

In general, NLP is used as a method for personal development through the "enhancement" of one's skills, which aims to have greater self-confidence and to communicate better with others.

The criticisms

To date, the efficacy of NLP has not yet been demonstrated, despite the fact that more than 40 years have passed since its conception, nor has this practice been the subject of rigorous scientific analysis, as happened for example for cognitive psychological therapy-behavioral. This has led to an absence of formal regulation, which can give rise to both arbitrary interpretations of the method and to its "manipulative" use.

Furthermore, scientific research on NLP has found contradictory evidence.

Some studies from the 1980s to 1990s have proven the benefits associated with NLP. For example, a study published in the journal Counselling and Psychotherapy Research found that patients in psychotherapy (various addresses) had improved symptoms and quality of life after being treated in association with the NLP methodology and psychotherapeutic treatments.

However, a review published in the 2014 British Journal of General Practice refutes the effectiveness of NLP with at least 10 studies. In summary, according to these studies, there is little evidence to assert that the method works with demonstrative evidence, especially when it intends to treat people's health, including mood disorders, weight management and substance abuse. According to scholars, the positive effects on the impact of the method would exist, but they are not exhaustive and not convincing. Only 18% of all research on NLP found clear cause-and-effect support for the theories underlying the method.

However, it should be noted that the research was conducted mainly in therapeutic contexts, and not in the commercial context, where NLP would find greater use.

One of the most used communication techniques in the last twenty years is Neuro Linguistic Programming. Known as NLP, it is one of the "sciences" most used by researchers, athletes, consultants, managers, training experts and professional communicators.

NLP was born and developed in California in the 70s, thanks to the collaboration of the mathematician Richard Bandler and the famous linguist John Grinden.

Reserved for a few in the past because of the high cost of the courses and the difficulty of the books that dealt with the subject, this science gradually manages to make its way into the psychological sphere.

But what exactly is NLP?

It is a psychological method that studies people's behavior, analyses models and thus extracts the practical techniques to teach to potentially overcome any situation (work, success, relationships).

What NLP teaches is that each of us can, with willpower, change and revolutionize one's life in an instant, abandoning limits through the help of concrete techniques. Each person is the architect of his own destiny, determined exclusively by our decisions and not by the living conditions as many believe and which are already "prescribed" and not changeable.

The message that NLP wants to convey to us is the secret of living well, that is " living life trying to make the most of it". It is we, and only us, who can make everything we want possible, starting from the determination, constancy and determination, which from the beginning must not be missing, together with the desire to fight for a purpose, therefore the energy that goes put into practice, up to the exercise, application, construction and achievement of our goal.

In a nutshell, this method would help us become the people we always wanted to be; an opportunity to learn how to use our mind and body in the most functional way possible.

Talking about NLP is equivalent to dealing with themes based on creativity, on freedom, on self-esteem, on choices and therefore on courage. The founders of NLP in fact coined this term (Neuro Linguistic Programming) precisely to highlight a link between neurological processes (neuro), language (linguistics) and the various behavioral screens that have been learned only through experience (programming).

It is indeed impossible, according to Bandler and Grinden, to find a field where this model cannot be applied: from self-esteem problems to sports or school skills, from courtship to success, and there are those who even claim that this discipline somehow manages to fight depression and other psychological disorders.

Summing up, NLP has among its main purposes, the goal of developing successful habits / reactions, amplifying effective behaviors to make what we want for us to happen and decreasing unwanted ones, which limit the occurrence of our design drawings.

With Neuro Linguistic Programming you learn to model the quality of the internal images lived and the sensations perceived so that they act for our benefit in the future. NLP makes us aware of our unconscious behaviors and programs that we can modify as we wish.

There are NLP academies where you can learn and put into practice all the possible techniques to achieve what you want a purpose: motivation, the basis for all our desire, is the ingredient fundamental that pushes us to fight to get it. Without it none of us would be able to reach the end to which it aspires.

Do you want to understand who you are facing in 60 seconds?

Do you want to sell successfully?

Would you like to increase your charisma or have a seductive and convincing voice?

According to the method in question, this can be done and much more.

Optimism, joy of life and cooperation are the three secrets of living in harmony. Everyone needs a paladin.

Have you ever wished for some things for yourself and done nothing to get what you claim you want? Have you ever wanted to lose weight, to free yourself from the slavery of some addiction such as alcohol or smoking, to want to learn a language or play an instrument?

Certainly yes, but how many of you have put your goals into practice and how many others have been stuck? This is how it happens, for all those belonging to the second sphere, who finds us doing only one action: complaining. The coach then takes over and helps us take stock, find our orientation, define ourselves as people. It spurs us to find a motivation in the event that our "journey" becomes tortuous.

There are four types of coaches:

LIFE COACH: the one who helps us achieve personal goals;

COMPANY COACH: the one who helps companies and professionals in the sector to act more effectively and with determination in professional life;

CAREER COACH: one who helps people in the phases of professional change, therefore a career jump or even a professional regression;

SPORTS COACH: helps students raise the level of performance and thus triggers mental and physical training.

The main objective of NLP is therefore to explain to us how everything we are is the simple result of what we have thought. Our life is in our hands.

Don't believe it? Try it for yourself!

NLP is an attitude ...

Characterized by a sense of curiosity, adventure and desire to learn the necessary skills to understand what types of communication affect others. It is the desire to know things that are worth knowing. It is looking at life as a rare opportunity to learn.

NLP is a methodology

based on the principle that every behavior has a structure ... and that this structure can be extrapolated, learned, taught and even changed. The guiding criterion of this method is to know what will be useful and effective.

NLP is a technology ...

which allows a person to organize information and perceptions in order to achieve results deemed impossible in the past.

Neuro-Linguistic Programming therefore deals with the study of the structure of subjective experience and what can be calculated from it.

His basic belief and promise are that effective thinking strategies can be identified, assumed and used by anyone who wishes to.

NLP was born from the fruit of years of research, carried out by Richard Bandler and John Grinder, aimed at finding out what the behavioral and linguistic elements were that allowed successful people to have such a significant constancy of positive results.

The results were the identification of a series of specific and reproducible behavioral strategies and linguistic models.

The hoaxes of neuro linguistic programming

From psychotherapy to coaching, NLP is still without solid experimental evidence, and has many characteristics of pseudo sciences.

Neuro-linguistic programming (NLP) has been rejected by science in every possible and imaginable way, yet it continues to be talked about.

The more attentive will have noticed that in the film Kingsman - Secret service (2014) neuro linguistic programming is passed off as a weapon of seduction.

It is in fact almost impossible to find a field where the NLP, according to its supporters, cannot be applied: from courtship to leadership, from self-esteem problems to sports skills, success is at hand, and there are even those who come to propose discipline to combat depression and other psychological disorders.

But what exactly is neuro-linguistic programming?

The origins

The NLP was born in the first half of the 70s, the golden age of the New-Age, and perhaps it is no coincidence that the crib was the California lysergic. The dads in the new discipline were Richard Bandler, a psychology student at the time, and linguist John Grinder, both from the University of California, Santa Cruz, who had begun to work out a sort of 'theory of everything' of psychotherapy from their respective fields. of study.

One of the cornerstones of the new, revolutionary branch of psychology would be imitation or, as the adepts define it, modelling: imitating the language and behavior of successful people it would be possible to make our skills our own and achieve their own results.

Chapter 18
Deception

The going with kind of mind control that will be investigated is deception. This mind control technique will have two or three likenesses' to control in the way those controllers will use a great deal of deception so as to locate a functional pace objective. This fragment will go into more bits of information concerning how deception functions, the methodologies related with it, and a piece of the examination that has been found.

What is Deception?

In any case is the definition about what deception is. Deception, alongside subterfuge, confusion, imagines, misleading, and beguilement, is a show used by the position to spread emotions in the subject about things that are contortions or which are basically almost the whole way feelings. Deception can consolidate a collection of things, for example, disguise, and spread, impedance, capable deception, presentation, and dissimulation. The director will have the decision to control the cerebrum of the subject considering the way that the subject will trust in them. The subject will recognize what the ace is communicating and may even be basing reachable plans and forming their reality dependent on the things that the expert has been letting them know.

On the off chance that the master is rehearsing the methodology of deception, the things they have been telling the subject will be counterfeit. Trust can without a considerable amount of a stretch be pummeled once the subject discovers, which is the clarification the ace must be gifted at the technique of deception and exceptional at getting something moving if they need to proceed with their subject.

Typically, deception will come up the degree that affiliations and it can incite sentiments of vulnerability and unfaithfulness between the two partners who are in the relationship. This is considering the way that deception hurts the rules of most affiliations and is in like way observed to impact the needs that go with that relationship. Considerable number people need to have the choice to have a real discussion with their embellishment; if they have discovered that their partner is surprising, they would need to understand how to use confusion and impedance to get the solid and reasonable data that they need. The trust would in like way be gone from the relationship, making it difficult to develop the relationship back to where it had once been. The subject would dependably be exploring the things that the ace was outlining for them, thinking about whether the story was authentic or something made up. Because of this new vulnerability, most affiliations will end once the subject finds a couple of arrangements concerning the deception of the master.

Sorts of Deception

Deception is a sort of correspondence that depends upon oversights and lies so as to persuade the subject of the world that best fits the ace. Since there is correspondence required, there will in like way be a few specific sorts of deception that could be happening. As appeared by the Interpersonal Deception Theory, there are 5 undeniable sorts of deception that are found. A piece of these have been appeared in different sorts of mind control, displaying that there can be some covering. The five basic sorts of deception include:

Deceptions: This is the place the overseer makes up data or gives data that is by no means equal to what is reality. They will demonstrate this data to the subject as truth and the subject will consider it to be reality. This can be risky since the subject won't grasp that they are being proceeded with sham data; if the subject comprehended the data was false, they would not likely be talking with the power and no deception would happen.

Avoidances: this is the place the head will make negating, crude, or degenerate clarifications. This is done to lead the subject to get disordered and to not get a handle on what's happening. It can correspondingly assist the head with disguising any trace of disappointment if the subject returns later and tries to reprimand them for the phony data.

Mask: This is one of the most for the most part saw sorts of deception that are used. Masks are the place the manager disregards data that is material or essential to the specific condition, deliberately, or they look into any immediate that would cover data that is fitting to the subject for that specific setting. The chairman won't have truly misdirected the subject; anyway they will have ensured that the basic data that is required never makes it to the subject.

Bending: this is the place the director will exaggerate a reality or distort a touch to turn the story the way wherein that they may require. While the authority may not be truly deceptive the subject, they are going to cause the circumstance to appear as though a more conspicuous strategy than it truly is or they may change reality a piece with the target that the subject will do what they need.

Under-depictions: a modest depiction of the truth is the exact opposite of the paltriness device in that the head will make light of or constrain bits of this present reality. They will tell the subject that an occasion isn't that goliath obviously of activity when in truth it could be what picks whether the subject finds the opportunity to graduate or gets that colossal progress. The ace will have the choice to return later and say how they didn't perceive how epic of a strategy it was, leaving them to look amazing and the subject to search in every way that really matters irrelevant on the off chance that they fight.

These are only a couple of the sorts of deception that may be found. The star of deception will use any procedure that is open to them to locate a useful pace objective, much like what happens in different sorts of mind control. If they can appear at their objective using another methodology against the subject, by then they will do it so the rundown above isn't the littlest piece specific. The chairman of deception can be staggeringly perilous because the subject won't have the choice to admit all with what is and what a demonstration of deception is; the star will be so talented at what they do that it will be essentially difficult to comprehend what is reality and what isn't.

Habits of thinking in Deception

Specialists have asserted that there are three essential objectives that are open in deceptions found in agreeable affiliations. These would unite frill focused points, pompous perspectives, and relationship focused habits of thinking.

We should take a gander at the embellishment focused desires first. At this moment technique for thinking, the manager will use deception so as to abstain from making hurt the subject, or their collaborator. They may in like way use the deception to ensure the subject's relationship with an outside outsider, to refuse having the subject stress over something, or to keep the assurance of the subject immaculate. A significant part of the time, this sort of inspiration for deception will be seen as socially important comparatively as socially mindful.

This sort of deception isn't as awful as a section of the others. On the off chance that the director gets some answers concerning something loathsome that the subject's closest companion said about them, the expert may close-lipped regarding it. While this is a sort of deception, it helps the subject keep that participation while protecting the subject from feeling frightful for themselves. This is the sort of deception that is discovered the most generally speaking seeing somebody and may in like way not cause that much harm whenever found. Most couples would use this sort of deception so as to ensure about their right hand.

Next is oneself focused way of thinking of deception. This one isn't viewed as respectable as the first and is right now looked slipping on than changed techniques. Rather than battling with the subject and how they are feeling, the executive is going to simply consider how they feel and about their own unique mental self portrayal. At the present time, chairman is using the deception so as to ensure about or improve their own psychological self view. This kind of deception is used so as to shield the star from assessment, mortification, or stun.

Right when this deception is used in the relationship, it is overall seen to be a more significant issue and offense than what is found with the frill focused deception. This is considering the way that the master is acting in a prejudiced manner as opposed to trying to ensure the relationship or the other right hand.

At last, the relationship focused technique for thinking of deception. This deception will be used by the chairman in the longing for keeping any shrewdness that may go to the relationship just by evading social injury and struggle. Subordinate upon the condition, this sort of deception will every now and then assistance the relationship and at different occasions it may be the clarification behind hurting the relationship since it will make things sensibly caught. For example, on the off chance that you cover how you are feeling about dinner since you would slant toward not to get in a battle, this may bolster the relationship. Then again, on the off chance that you busy with extramarital relations and close-lipped regarding this data, it is basically going to make things progressively confused at long last.

Notwithstanding the purpose of deception in the relationship, it isn't proposed. The ace is holding data that may be fundamental to the subject; when the subject finds a couple of arrangements concerning it, they will begin to lose trust in the ace and marvel what else the executive is escaping from them. The subject won't be unreasonably worried for the clarification for the deception, they will fundamentally be vexed that something has been kept from them and the relationship will start to have a section. It is once in a while best to stay with the arrangement of validity in the relationship and encircle yourself with people who don't rehearse deception in your social party.

Perceiving Deception

If the subject is enthused about maintaining a strategic distance from deception in their life so as to keep up a key decent ways from the mind games that go with it, it is as frequently as conceivable a sharp plan to understand how to recognize when deception is going on. Reliably, it is difficult for the subject to find that deception is going on except for if the master goofs and either lies that is clear or noticeable or they repudiate something that the subject undeniably knows to be genuine. While it might be difficult for the chairman to cheat the subject for an important stretch, it is something that will by and large happen in typical ordinary nearness between people who know one another. Recognizing when deception happens is regularly hazardous considering the path that there are less any pointers that are completely solid to tell when deception occurs.

Chapter 19
Mind Games

When a person plays "mind games" on us, it is attributed to being innocent. Many people have come across this at some point in their life. Take an example when someone is planning a surprise party and doesn't want the other person to know and he does this by playing mind tricks in order not to give away what the surprise actually is. This is merely considered innocent and silly. Dark psychology mind games are not in any way innocent. Mind games in dark psychology are attributed to the hypnotist toying with the will power and sanity of his victim. This differs from other dark psychological manipulation in the sense that the manipulator is playing with his victim for his own pleasure and enjoyment and is not invested in what the outcome will be. His interest in the victim would be to test the victim so to speak. Mind games are used by a hypnotist when other forms of suggestions to the victim are not effective and may decide to use mind games which are rather less obvious to the audience. The manipulator may decide to use mind games to his own pleasure and amusement. Mind games are very effective in reducing the assuredness and psychological strength of the victim. The victim is eluded into thinking that he still has control. Manipulators are able to satisfy their twisted amusement when playing mind games. Such dark psychological manipulators do not see their victims as equal human beings and instead chooses to see the victim as a 'toy' and a person who can be manipulated and therefore, watch with amusement when victims do what they tell them to. Sometimes, a dark manipulator will have known mind games all his life and knows no other forms of dark psychology manipulation. These manipulators can be dangerous because they know not of any other option and therefore no need of changing and being more humane. Let us dive into the specific types of mind games used by dark manipulators.

Ultimatum

An ultimatum can be defined as a final proposition or condition. One, therefore, is presented with a severe choice. They are viewed more as demands other than a request. An example is, "Be more outgoing…or I will see other people". Certain factors will decide whether an ultimatum will be considered as a mind game. The three factors are one, the type of person giving the ultimatum, second the intention for giving the ultimatum and lastly the nature of the ultimatum.

Persons who give ultimatums and genuinely care about the persons and have a valid reason for doing so, and then it will fall under the non-dark manipulation. These persons will generally include spouses, parents, siblings or close relatives. However, if they fall into any of the categories mention it does not necessarily rid them of dark intentions from the ultimatum given.

What was the intention of the person giving the ultimatum? People with good intentions are often driven by the desire to help or assist in bettering the life of a person. Where a person gives an ultimatum to for example stop smoking or drinking too much, then this seen as good intentions. Being able to tell the intention of an ultimatum is difficult and so looking at the nature of the ultimatum itself is the surest way to be able to tell whether it is dark.

Dark manipulative ultimatums will involve the person doing something that goes against what they stand for and goes against what their self-interest. The victim ends up comprising their moral standards in the process. Manipulators test their victims to see how far they go in compromising what they believe in. As we have seen, non-dark ultimatums are usually to benefit another person and the does not have to go against what they know is wrong.

What is a dark psychological ultimatum? The person giving the ultimatum will be a friend, a boss or a person who the victim is in a toxic relationship with. It could also the form of a spouse, a parent or a sibling. The manipulator will often give ultimatums that go against the victim's moral conviction or that which can possibly be dangerous to the victim. Here, the dark manipulator will notice a disinclination towards something and take advantage of this to make their victim do their bidding. An example will be a girl who is not comfortable in wearing costumes or revealing clothes. Some of the ultimatums will be, "It's an only costume party, it is either you wear one or you are not invited". Some ultimatums lead to harm to others such as assault and even murder. At very extreme cases, the victim ends up taking his own life in completing a suicide pact in which the manipulator does not honour his end.

The External Break up

Everybody likes to be in a relationship where there is that sense of security and knowing that your partner is content. A manipulator will know this but will use these for their dark intentions. A manipulator will ensure that their partner will be powerless by instigating feelings of instability, and negativity within their relationship. This technique of 'The External Break up' is often deployed in a romantic relationship. It manifests itself when a partner continuously to scares the other that he or she will leave them. This is aimed at creating feelings of anxiety and instability within the relationship. This mind game takes the form of promised breakups, implied breakups and actual breakups that do not happen.

Implied breakups are those that are not expressly stating the words 'break up'. Instead, the manipulator throws hints there and then to create some doubt in the partner's mind. They can do this by making statements that exclude their partner from future plans together. Promised breakups happen where the dark manipulator scares their partner that they instead to break up with them somewhere in the near future. Words like, "Don't worry I won't have to deal with this anymore because I'll be leaving soon" show the intention of a breakup in the future. Promise breakups fall in between the implied breakups and the actual breakups. Where the dark manipulator mentions the idea of cutting ties with their partner, either by divorcing, separating or breaking up, but does not follow through then it calls under the promised breakup.

The actual break is the most severe compared to the implied and promised breakups. It happens when the manipulator decides to leave their victim without actually leaving in the end. They may pack up their clothes and belongings in the attempt to leave but once they see the sadness all over their victim's face, the decide otherwise.

After going through and understanding the tactic of the "external break up" we ask ourselves what therefore is the end game for manipulator when they use this tactic? The manipulator aims at having the upper hand in the relationship by creating feelings of uncertainty and lack of security from the life of the victim and therefore reducing their power in the hands of the manipulator. By repeatedly simulating a breakup with the victim, the manipulator is trying to test the waters of how far one will go in putting up with being treated like a toy. In the end, when the manipulator gives in to the victims' begging for the relationship to continue, they make themselves look like the generous ones. This works so well for the manipulator because his or her victim is not thinking rationally to be able to figure out why they relationship should end. They are therefore willing to continues with the relationship. Many people do not understand this concept of dark psychology and why a person would want to continue to be in a relationship with a dark manipulator in the first place. The impact of this on the victim includes the likelihood of developing serious trust issues where they will have a hard time trusting another person. This could take a toll on the victim's professional relationships and family relationships as well. After a long period of constant threats, the victims become almost like a slave to the manipulator in which the manipulator eventually grows tired and moves on to their next prey.

Hard to get

And just like ultimatums, the hard to get tactic can easily pass off as being normal. Hard to get can be dark as it can be also harmless and normal. Hard to get when it is harmless it occurs when a person will want to make them seem trying to be with them is not as easy. They will do this by making themselves less available by not making to every date and leaving the phone to ring a couple of time before finally picking up. The 'hard to get' dark psychology is much riskier. The manipulator will use this tactic during the relationship rather than at the beginning of the relationship. Unlike the innocent hard to get where the intention is to eventually be in a happy relationship, dark psychology hard is far from taking into account the wellbeing of the victim. When used at the beginning of the relationship it is innocent because no expectations are infringed at this point. At this point, no one is dependent or reliant on either of the person, so no harm comes from playing hard to get. Further along in a relationship when things are going on well then suddenly a person is unreliable and often times tries to make themselves busy. This kind of behaviour is not normal because relationships are about making and spending time with each other as this will firm up the relationship. A manipulator will be very cunning and start pulling away when their partner us already reliant on them. The victim will therefore put an extra effort to reconnect with their partner. In the end, the manipulator has the upper hand and will use this power to his or her own purpose while the victim is left in deep confusion and instability.

Chapter 20

Persuasion vs. Manipulation

The line between persuasion and manipulation is so thin that it often gets blurry. Distinguishing these two concepts can often be difficult, especially depending on the circumstances and your own perspective as an individual. Persuasion and manipulation are alike in that in both cases there is someone trying to influence the decisions and behaviors of another. The key distinction between the two is that manipulation is seen to be highly driven by self-interest where one party is willing to go through any length to benefit themselves, including putting others in harm's way. Persuasion on the other is the nicer cousin of manipulation--there is a desire to influence for self-interest but there is often a line drawn to mark boundaries. Persuasion is the more ethical way to go about it, many will argue. When all's said and done, however, the two concepts seem to intertwine especially depending on the techniques used to achieve either of them.

People always have different ideas of what words mean, but to be successful in manipulation and persuasion, you need to know the different ways these terms are understood as well as what we mean when using them in this book. In common speech, persuasion is considered a neutral word; of course, someone can be persuaded to do something that helps the persuader and not themselves, but the word itself does not imply that. Manipulation, on the other hand, tends to mean ill intention of the manipulator.

The ethics of manipulation and persuasion are a topic we have explored throughout these pages, but know that for our purposes, persuasion is changing someone's beliefs, while manipulation is changing someone's actions. This is easy to remember, because NLP involves the neural pathways for both language (belief) and programming (action).

If you want your subject to change their behavior, you have to get them to change their thinking about their behavior. They are a thinking person just like you are, and while they have mental shortcuts that can get in the way (just like they can for you), your subject is entirely capable of talking through their judgment calls with you. In a conversation with you, they can come to re-evaluate their actions, and if you go through the conversation the right way, you will have the opportunity to convince them to change.

When it comes to manipulation, there is a slight difference from persuasion. The difference is that at some points, it is, in fact, the right thing to ask them to change their behavior directly. Now, you don't want to pull this out as your first move. This is something you build up to after a long conversation — after you accomplish steps zero and one, just as you do for persuasion. But the big difference between changing someone's ideas and changing their behaviors is here in step two: more often than not, you should directly tell them what you think they should do differently.

When NLP newcomers learn this at first, they are totally taken aback. They think, how could I possibly be told to tell them directly to change their behavior? But if you think through it a little longer, it makes sense. What is the difference between belief and behavior? Persuasion changes belief by getting close to someone's mind and changing what is in there, and manipulation is getting closer to their mind and changing what is in there, too.

But with manipulation, there is the added hurdle of getting them to follow up on the change in thought. While it is absolutely true that all of our behavior ultimately comes from our mind, our brains are still not simple masters of our actions. Rather, our actions are determined by multiple factors other than simply what our brain tells us to do. The reason you eventually have to ask your subject to change their actions directly is that for new behaviors, a change in thought is just not enough.

Your subject needs voices other than the one in their head, telling them what to do. They have the thought you got into their head through NLP; you are telling them directly, too. But there is still more you have to do.

Social bonds are incredibly important to human beings. If you want to manipulate someone's behavior, unlike when you persuade them into having new thoughts, these thoughts alone are not enough. You telling them what to do is not enough, even once they have recognized you as like them. If you want to change their behavior, next, you have to change the social environment of the person with the undesired behavior.

This is not a catch-all for manipulation, because nothing is. After all, not in every situation will you be able to change the social environment of your subject. If they are not friends or family, but rather a co-worker, this could prove much more challenging. It is only fitting since manipulation is a more difficult and complicated task than persuasion.

But if this is a person whose social environment you have some control over, you have to determine what social factors are leading to undesired behaviors. Is there another family member enabling their drinking or drug use? This is the most prominent example, but all of it is emblematic of the NLP manipulation framework in general.

All of this is to say that when you are not in control of a person's social environment, directly telling them what action they should take is a necessary and challenging part of the process. It is so challenging because there is no way around it, and it is also very easy to do the wrong way.

You have to work hard not to work too hard for them. If they can see how badly you want them to change their behavior, they will want to continue acting the way they do out of spite. Don't give them this opportunity.

Recall how with persuasion, we said never to address objections to your frame. In fact, if at all possible, you don't want to address the frame itself. That's because if you address the frame itself, you are acknowledging the fact that it is not the naturally-occurring reality that you want your subject to see it as. However, with manipulation, the situation is different than it is for persuasion.

With manipulation, you have to respond to objections directly, because you have to tug harder than you do with persuasion. You see, persuasion is a subtler, quieter art than manipulation. This is a direct way of saying that manipulation is loud and aggressive, because it is not.

But you can't be quite as gentle with manipulation. You want them to change their habits, so in order to get your subject to understand the gravity of the situation enough to trigger the behavior change, it is necessary that you are slightly pushier than you are with persuasion. Again: don't be pushy, but you can't be as subtle as you are with persuasion.

Even when you deal with their objections, you are better off preparing for them before they come up. When you are ready for any question or complaint your subject can haul at you, it is a signal to them that you are like them, you see things from their side, and perhaps, you know better. This is Step One yet again. If you demonstrate that you are like them and can reason things out better, they will listen. You are almost ready to get into the techniques of manipulation, but before then, you need to get into the personality of the NLP manipulator.

You might think that you are born with a certain personality, and you can't do anything to change it, but this couldn't be further from the truth. In fact, the kind of personality you should adopt to get people to do what you want is one that anyone can learn.

Why is learning this personality so important? Well, it's important because you need to seem like you are positive about what you are saying. If you seem even a teensy bit unsure in any of your speech or your body language, nobody is going to buy what you are selling. That's why in your body language, dress, facial expression, tone of voice, and words, you need to pull off the personality of someone who knows what they are talking about.

They have the answer to your question; they know what's what. If you can pull off that personality, you basically don't have to do anything else. Personality is everything — don't forget that.

Personality is so important because no matter how unlikely something seems on the surface, if it comes out of the mouth of the right personality, people will believe in it. You have to believe in what you are saying to some extent if you expect to pull this off, so don't think you can playact your way through the whole thing — after all, you are not doing the personality right if you are unsure about the merits of what you are saying. But more important than anything you say is the personality you are displaying while you say it.

Not everyone has this naturally, but it is not nearly as hard to learn as you might think. The right place to start is always your breathing and your posture. You already know what the right posture to take is — stand up straight and without shaking. Now, take deep breaths like for your state control exercises. Just like before, don't breathe loudly. Breathe deeply but not in a way that anyone can tell is unusual.

The third and final thing you have to do is enter the headspace of this unshakeable personality. Everyone has experienced a moment where everything was going right for them, and that is exactly the place you need to go. Revisit that memory as though you were there again right now, and come back as the person you were in that memory.

The world is at your fingertips just like it was back then, especially if you carry this person inside of you. That person is necessary to succeed in manipulating people's behavior in the techniques coming up, so be sure you have your personality ready before reading. You won't be able to pull these off otherwise.

Chapter 21
Mirroring

First things first, you must learn what mirroring is. At the simplest, it is the human tendency to mirror what is happening around them when they feel a relationship to whatever it is that is around them. For example, if you look at an old married couple, they are likely to constantly be mirroring each other's behaviors. It is essentially the ultimate culmination of empathy—the individuals are so bonded, so aware of each other and their behaviors, that they unconsciously mimic any behaviors that their partner does first. The two married people at the diner may both sip at their coffees at the same time as each other, or if one drinks, the other will follow shortly after. If one shifts in his seat, she will do so as well, always leaning to mirror the position her husband is in. If she brushes off something on her shoulder, he will unconsciously touch his shoulder as well. This act is known as mirroring, and it occurs in a wide range of circumstances.

You do not necessarily have to be a married couple that has been together for decades for mirroring to be relevant, either—you can see it everywhere. The person interviewing you for a job may begin to mirror you when the interview is going well, or the person who thinks that you are attractive may mimic some of your behaviors as well. You can see these behaviors mimicked started quite early on in terms of how long people have been interacting as well—sometimes people will even hit it off right off the bat and begin mirroring each other, emphasizing the fact that they seemed to have clicked.

Mirroring is essentially the ultimate form of flattery—it involves literally copying the other person because you like or love them so much. Children mirror their parents when learning how to behave in the world. Good friends often mirror each other. Salespeople wanting to win rapport, mirror people. No matter what the relationship is, if it is a positive one, there are likely mirroring behaviors, whether unconscious or not.

Uses of Mirroring

You may be wondering why something as simple as mimicry can actually be important to others, but it is actually one of the most fundamental parts of influence, persuasion, and manipulation. When you mirror someone, you can develop rapport. Rapport is essentially the measurement of your relationship with someone—if you have a good rapport with someone, you have developed some level of trust with them. The other person is likely to believe what you are saying if you develop rapport. However, if you have not yet developed rapport yet and you need the other person to listen to you, you can oftentimes artificially create that rapport through one simple task—mirroring. If you mirror the other person, you can essentially convince him to develop a rapport with you, whether it was something he wanted to develop on his own or whether you forced the point.

By constantly mirroring the other person, you essentially send the signs to their brain that they need to like this person because this person is just like them. Remember the three key factors for likability? The first one was able to relate or identify with the other person. In this case, you are presenting yourself as easy to relate to simply because you want the other person to like you. With liking you comes rapport. With rapport comes trust, which you can use to convince the other person to buy cars, or do certain things that will benefit you. Building rapport even builds up the ability to be able to manipulate the other person—you need to be trustworthy for the other person to let you close enough to manipulate in the first place.

How to Mirror

Luckily for you, mirroring is quite easy to learn how to do. While it may seem awkward and unnatural at first, the more you practice it, the more natural it will become to you, and the more effective you can get at it. Remember, if you want to mirror someone, you will need to toe the line between too much and not enough. If you are too overt, the other person will catch on and will likely be more put off than convinced to like you. Take a look at these four steps so you can learn to mirror for yourself.

Build up a Connection

The first step when you are attempting to mirror someone is to start by building a connection somehow. If you do not feel the connection with the other person, they are not likely to be feeling a connection either. Keeping that in mind, you should begin to foster some sort of connection and rapport. This can be done with four simple steps on its own.

- Fronting: This is the act of facing the other person entirely. You start with your body oriented toward them, directly facing the other person to give them your complete attention.

- Eye contact: This is the tricky part—when you are making eye contact, you need to make sure that you get the right amount.

- The triple nod: This does two things—it encourages the other person to keep speaking because the other person feels valued and listened to, and it makes the other person feel like you agree with them. It develops what is known as a yes set. The more you say yes, the more likely you are to develop a connection with the other person.

- Fake it till you make it: At this point, you have spent a lot of time setting up the connection, and it is time for the moment of truth. You should imagine that the person is the most interesting in the world at that particular moment. You want to really believe that they are interesting to you. Then stop pretending—you should feel that they are actually interesting to you at this point. This is the birth of the connection you had been trying to establish.

Pace and Volume

Now, before you start mimicking their body language, start by paying attention to the other person's vocal cues. You want to make sure you are speaking at the same speed as the other person. If they are a quick speaker, you should also speak quickly, and if they are a slower speaker, you should slow your own speaking pace down to match. From there, make sure you are also mimicking the volume. If they are louder, you should raise your own voice. If they are keeping their voice down, you should follow suit. These vocal cues are far easier to mimic undetected than the rest of the physical cues.

The Punctuator

Everyone has a punctuator they use for emphasis. It could be something like a hand gesture that is used every time they want to emphasize something, or it could be the way they raise their brows as they say the word they want to stress. No matter what the punctuator is, you should identify what it is and seek to mimic it at the moment. Now, oftentimes, this cue is entirely unconscious on the other person's part, and as you begin to mimic it, the other person is likely to believe that you are on the same wavelength. This should really do it for you without making what you are doing obviously.

The Moment of Truth

Now, you are ready to test whether you have successfully built up the rapport you need. When you want to know if the other person has officially been connected to you, you should make some small action that is unrelated to what you are doing at that particular moment and see if the other person does it back. For example, if you are having a conversation about computers, you may reach up and rub your forehead for a split second. Watch and see if the other person also rubs at their forehead right after you. If they do, they have connected to you, and you can begin to move forward with your persuasive techniques.

What is the Barnum Effect?

The Barnum effect is the effect that you see when people take something exceedingly vague and declare that it must be tailored to them. For example, imagine a horoscope—People often talk about how much of a Taurus they are because they are so stubborn, practical, and ambitious. Never mind the fact that many people can describe themselves as stubborn, practical, and ambitious, the people are wholeheartedly convinced that those vague descriptions of a person's personality are so specific that they must be trusted. This concept applies to many different paranormal instances, such as astrology, as briefly touched upon, and fortune-telling.

People will fall for vague hints at something that is clearly fishing for feedback in order to get something that could actually be utilized in a way that would be beneficial for the one attempting to manipulate others. People think that even the vaguest of hints is enough proof to legitimize whatever is being said, so long as they can at least in part identify with it.

Chapter 22

Seduction and Dark Psychology

Seduction is persuading someone to have sex with you or make them more excited to do so. Seduction is often simply part of attraction between two people, as it sets the stage for the sex to come. It may include a woman wearing lingerie to greet her partner after his long day at work, or a man buying his date a fancy dinner and whispering in her ear how beautiful she looks in her dress. Among good-intentioned people, seduction is a normal, specific mode of communication appropriate for indicating desire and hoping the person of interest feels the same way. Ideally, the process of seduction is never dishonest or misleading. The person being seduced knows what their pursuer wants, and they can have a mutually satisfying sexual encounter or start a romantic relationship.

How does seduction fit into dark psychology then? Seduction can be to help the person being seduced, to hurt that person, or to benefit the person doing the seducing. All three of these motives have one thing in common though; all seduction, to some degree or another, requires at least some affection of the desired person's mental state.

Why Use Dark Seduction?

Dark psychology seduction is often an effective seduction technique because it can make the person of desire feel intrigued and excited. In some ways, it is almost a form of persuasion. At its most ethical, persuasion is beneficial to the person being seduced and the persuader has good intentions. At its worst, the persuader causes harm to their victim and only thinks of possible rewards to themselves. The same goes for seduction.

A person with good intentions may use dark psychology seduction techniques gets the most out of their love life. They harbor no will to harm others but know how to have fun. When this person decides to marry, it will most likely be a happy marriage, as they have created excitement and joy in their partner.

Someone who seeks to harm through dark psychology may choose to do so because of the thrill they derive of letting someone down so spectacularly. When this harmful person seduces someone, usually a vulnerable person, they feel pleasure in watching the partner's excitement turn into fear and anguish. This person has no regard for the person they have seduced and is often promiscuous, with a long trial of failed relationships and angry exes behind them.

If someone is completely self-serving with their dark seduction, then their results will fall somewhere between those of the good intentioned seducer and of the maniacal seducer. The completely self-serving seducer may cause harm but mostly out of selfishness and lack of awareness. In general, this person will often be dissatisfied because their intentions lead them to neglect their relationships with those whom they seduce.

Using some dark psychology when trying to seduce someone is not innately good or evil. Instead, it simply a tactic that has proven more successful than others. The only absolute truth about this method is that it is an efficient way to seduce someone, that maximizes one's chances of finding someone they find attractive and enjoy. With that said, dark seducers are more likely to get what they want because they know what they want. Dark seducers usually get the most attractive, most successful partners because they see what they want and go for it. They are not wishy-washy, and they do not settle out of convenience or loneliness.

Dark Seduction Techniques

There are many techniques to dark seduction, but at the core of this method is creating excitement and joy in whomever you wish to seduce. Be sure to entice this person, make them want you. These techniques are all about creating witty banter and showing off how fun and attractive you are in a suave, smooth, swoop.

The Friendly Opener

This technique involves doing anything but asking "what's your sign" or "come here often?". In this technique, an open-ended question is best. Something like, "Hey there, could you help my friend and I? We have a disagreement over who the most overplayed artist on the radio is right now." See what's happening? The seduce asked a friendly question that opened the floodgates for a funny, friendly conversation. The person being seduced does not feel overpowered or intruded upon. Instead, a friendly stranger asked an interesting question.

The strength of this technique is that it simply invites friendly conversation without mentioning any sex. It is impossible to be rejected, because this is not even a sexual advance. It is simply a way of meeting a new person and having some fun banter.

This also works because it avoids coming on too strong. The object of your desires is less likely to feel defensive, suspicious, or intruded upon if you present yourself in a friendly, non-aggressive way. You will not seem to overtly sexual or creepy, so this person will not feel the need to avoid you or shut down conversation as soon as it ends.

Show Off (A Little Bit)

This tactic is all about demonstrating, not just bragging about, social capital and success. The first step is to simply look the part—wear a nice, noticeable watch or jewelry. Dress unlike everyone else, as it is a sign of a confident, independent thinker. Another important way to show off is not to seem too desperate for company; show up to a bar with a group of friends, or flirt with two women at a time who are friends with each other. This demonstrates to both women that you are not only a bit of a challenge but also that you are a friendly, confident person in general.

Be Mean (But Again, Just A Little Bit)

If the person you are trying to seduce is acting a little bit haughty or clearly playing hard to get, pretend you are about to walk away. They will be wowed by this one because they expected you too keep playing along with their little game. Showing that you are not so desperate that you will put up with game playing from them show you are confident and not in need of their attention because you can simply go seek attention elsewhere. Once they want your attention back, you have practically won, because now the person you are interested feels like they are working a little bit extra for your attention, not the other way around.

Send Mixed Signals

It can work in your favor not to seem too interested. Do not pick up their call every now and then. Maybe once whomever you are interested in seems to reciprocate, hold back just a little bit and cut back contact. Why? Seeming just a tad aloof can create a sense of depth about you, leading your target to wonder even more about you or maybe even become fascinated. This is all about creating an air of value about yourself. Everyone wants what they can't have. Playing just a little bit hard to get can make someone's interest in you pique.

Give The Ego A Nice, Long Stroke

To stroke someone's ego, do not simply flatter them until you turn blue in the face. Instead, agree with them a lot—go along with what they say, get to know them and come to understand how this person's emotions work. In doing this, you will comply with the person's belief that they are the main character of their own story. Everyone believes this about themselves because it is true; each person, in their own story of their life, is the main character of that story. By playing along with someone else's story of their own life, you satisfy and validate them, making them trust you and enjoy your presence.

Be A Little Bit Taboo

Most people are, at least to some extent, thrill seeking. This does not mean that we all seek out dangerous situations or abuse hard drugs to feel alive, but rather that we all crave a little bit of excite, and the taboo inspires this in us. Another important facet of dark psychology is to know that you miss one hundred percent of the shots you do not take. Why is this important? A little bit of arrogance and psychopathy is useful here. Most people suffer from an overwhelming fear of rejection when they are trying to flirt. They are so afraid in fact, that they will avoiding going after those they find attractive because they fear how much pain they may feel as a result of rejection. The dark seducer knows a secret, which is that there are billions of fish in the sea, and rejection is not all that bad. Simply put, learn to take rejection—avoiding it simply makes flirting an even more daunting prospect and adds even more anxiety to dating. Instead, flirt enough that you get some practice with rejection. Once you survive it a few times, it will seem way less intimidating. The dark seducer knows a rejection is a blessing in disguise, because it simply frees up time in the future to pursue other, more interested prospects.

Rejection creates resilience and will let you figure out what your flirting and seduction style may be. Some people go for a more structured approach, by asking questions and knowing the emotions to evoke in a specific order in the object of their desires. Others, however, may like to go for it more organically by asking an open-ended question or "going with flow"—of course, projecting confidence and ease throughout the interaction.

Chapter 23
How to Protect Yourself Against Emotional Predators

Dealing with Manipulators

You can come across manipulators in all aspects of your life, both professional and personal. Whether you want to believe it or not, even those you love the most and hold dear can be manipulators. You might have to deal with manipulative partners, manipulative parents, or even manipulative coworkers. Regardless of the manipulator, you are dealing with; you can use the tips given in this segment to deal with manipulation and manipulative people. It isn't always easy, but you must learn to do so. After all, you are the only one who is responsible for your overall wellbeing.

Basic Fundamental Rights

A fundamental right is inalienable, and no one can take it away from you. This is one thing you must keep in mind whenever you come across any person who is a psychological manipulator. You must not only recognize your rights but must also prevent the violation of these rights. As long as you don't harm others, you must stand up for yourself and protect your rights at all costs. If you knowingly harm someone, you may lose some of these fundamental rights. Here are a couple of basic human rights you must be aware of.

- You have the right to be treated with dignity and respect.

- You are free to express your opinions, feelings, desires, and wants.

- You are free to set your priorities, and no one can force you to do something.

- You don't have to feel guilty when you say "no."

- You have the right to set specific boundaries for yourself.

- You have the right to have different opinions, and you don't have to agree with everyone.

- You not only have a right, but an obligation to safeguard yourself mentally, emotionally, and physically.

All these fundamental rights define your boundaries. You must not only enforce your limitations on others but must also respect them yourself. Of course, you'll come across people who don't respect your rights. Especially those who resort to psychological manipulation, strive to deprive others of their rights so that they can exert control over you. However, keep in mind, you have the power to decide what you want to do, and you are the only one in charge of your life.

Maintain Some Distance

A manipulator often puts up a façade for the world to see and doesn't let his true intentions rise to the surface. A simple way to detect or spot a manipulator is to see the way he acts in front of different people and various situations. Most of us tend to exhibit social differentiation to a certain degree; emotional predators and psychological manipulators tend to dwell on the extreme ends of the spectrum. An emotional manipulator can be extremely polite one instant and unnervingly hostile the following. If you notice this kind of behavior from anyone in your circle, maintain your distance. If you cannot get away from such a person or avoid social interactions, then limit your interactions. Spend as little time dealing with such a person as possible. Even being around them will hurt you in ways you cannot begin to comprehend. You don't have to worry about being responsible for their feelings. If the manipulator tries to make you feel guilty for maintaining your distance, it is a part of his manipulative nature, and you're not obligated to fix them. So, stay away.

No Personalization

A manipulator is continuously going to look for your weaknesses, and once he understands them, he will exploit them. Therefore, he might try to make you feel inadequate, doubt your sanity, and question your judgment. If you experience any of these feelings, then it means the manipulator has a stronghold over you. Don't ever blame yourself in such situations because it only increases the power the manipulator has. In such instances, remind yourself, you are not the problem, and there is nothing wrong with you. Take a moment to think about the relationship you share with the manipulator and answer the following questions.

- Does this person seem to have unreasonable demands and expectations from me?

- Does he treat me with the respect I deserve?
- Is this relationship well-balanced, or does it only favor him?
- Does this relationship make me feel good about myself?

If your answer is in the affirmative, then there is nothing wrong with the relationship. However, if it isn't, then you are in a relationship with a manipulator. Your answers to these questions will give insight into the kind of person you're dealing with. So, stop blaming yourself, and instead look at the other person.

Probing Questions

A psychological manipulator will inevitably start making requests. These requests are subtly veiled demands. Often the claims made will be such that you are required to go out of your way to meet his needs. If the claim you're presented with seems to be unreasonable, it's time to shift the attention back onto the manipulator by asking a couple of questions. By doing this, you can judge for yourself whether the person has sufficient self-awareness to realize the unreasonableness of his demands. Here are a couple of probing questions you can ask.

- Is this a request or a demand?
- What will I get if I fulfill this?
- Does this sound fair to you?
- Does this seem reasonable?
- Do you expect me to (restate the demand) do this?

By asking such probing questions, you are placing a mirror in front of the manipulator to check his true nature and intentions. If the manipulator has even a little self-awareness, he will quickly withdraw his demand or even apologize for it. However, it is quite unlikely that an emotional predator will have any awareness about the unreasonableness of his request and might expect you to comply regardless. If the manipulator tries to turn the tables on you and say you are overreacting or are being unreasonable, steer clear of him. Either way, you have your answer.

Time Is Your Ally

Not only will the manipulator make unreasonable demands but will also expect an immediate answer. By doing this, he is trying to maximize the stress placed on you to exert a higher degree of control over you and on the situation. In such instances, don't play right into the manipulator's trap and buy yourself some time. A suitable response is, "I will get back to you soon," or "I will need to think about it." If you don't respond to this demand immediately, you are preventing him from controlling you. Once you have sufficient time, and can carefully analyze the situation along with its pros and cons. If you feel like it is an unreasonable demand, then you have the right to say "no."

Saying "NO"

A lot of people often struggle with saying "no." You must not only be firm while declining a request but must also do it diplomatically. After all, you do want to prevent the manipulator from creating an unnecessary scene, don't you? You have the right to say "no," and don't let anyone take this away from you. If you allow someone else to control your actions like a puppeteer, you are giving away your power to choose. You can say "no" whenever you want to, and you don't have to feel guilty about it. Don't let the manipulator shame you or make you feel guilty for not complying with his demands.

Confrontation

An emotional predator, like a manipulator or a narcissist, is essentially a bully. While dealing with a bully, keep in mind that they are often targeting those whom they perceive to be weak or soft targets. As long as you don't take any action, stay compliant, and passive, the bully will always have some control over you. A lot of bullies put up a facade of courage and are often cowardly on the inside. So, once a target starts disobeying them or not complying with their request, bullies tend to back down. This stands right not just for a bully in school, but also in a personal or professional environment. If you ever decide to confront a bully, ensure that you are in a safe and secure environment. Make sure the bully cannot harm you and if required, opt for public confrontations. Having a couple of witnesses around you will be quite helpful. If you need help, ask for it and don't try to do everything by yourself.

Importance of Consequences

You must not only establish certain boundaries but must also set consequences for the violation of those boundaries. Whenever you feel like someone is violating your limitations, you must deploy a result. This is an important skill, especially while dealing with tricky and unscrupulous individuals.

At times, regardless of all that you do, being around a manipulator can cause irreparable damage to your overall wellbeing. In such instances, you might have to sever all ties and run in the opposite direction. If that's what you need to do for your wellbeing, then don't hesitate. You owe it to yourself, and you deserve better than being manipulated. So, don't sell yourself short and don't subject yourself to manipulative abuse.

Chapter 24
Speed Reading People

What Is Speeding People?

Ignite the Art of Reading People through Your Super Senses

If you want to read people, you have to don the garment of a psychiatrist who has the power to interpret cues which are verbal and nonverbal. You need to observe beyond people's masks into their real self. You may not get the entire picture about anybody through logic alone. You have to surrender to their critical forms of information to interpret the essential nonverbal perceptive cues that individuals exude. For you to achieve this feat, you need to be eager to surrender emotional baggage like ego clashes or old resentments and also any preconceptions which can prevent you from making out the person. It is crucial, as well, for you to obtain information without bias and continue to be impartial without twisting it.

In the process of reading a colleague, your boss, or partner for you to understand them accurately, some walls need to come down, and you need to surrender biases. You need to be ready to let go of limiting, old ideas as far as intellect is concerned. Those who read other people well are taught to comprehend the hidden. They have discovered how they will draw on what is called 'super-sense' so they can take a profound observation beyond where you usually steer your focus when you attempt to hack into transformative awareness.

Examine cues of body language

When you are reading the cues of body language, you have to surrender the focus by releasing your struggle to understand the hidden signals of body language. Never get analytical or overtly intense. Stay fluid and relaxed. Observe by sitting back comfortably.

Focus on appearance

When you are reading other people, take note of what they are wearing. Are they putting on well-shined shoes and power suit? The indication for success is when someone deck out decently. For someone wearing a T-shirt and jeans may be an indicator of that person being comfortable with casual. It may be a signal of a seductive choice when someone wears a tight top with cleavage. A pendant like Buddha or cross may indicate spiritual values.

Notice posture

Postures are an essential aspect of reading people. It's a sign of confident when people's head is held high. Or you can get an indication of low self-esteem when they cower, or they walk irresolutely. You can also get a sign of a big ego when they have puffed-out chest and swagger.

Pay attention to physical movements

When you read others, look out for their distance and learning. In general, people bend forward at those they like and keep a distance from others they don't. Also, when people cross their arms and legs, you can see signs of anger, self-protection, or defensiveness. It is an indication that people are hiding something when they hide their hands by placing them in their pockets, laps, or place them behind them. With cuticle picking or lip biting, you will get a sign of people attempting to calm themselves in a difficult circumstance or under pressure.

Read facial expression

Our faces provide the outline for our emotions. Profound frown lines indicate over-thinking or worry. The smile lines of delight are crow's feet; pursed lips are a signal of contempt, anger, or bitterness. While teeth grinding and clenched jaw are indicators of tension.

Take note to your intuition

It is possible to tune into someone ahead of their words and body language. Though not what your head says, what your gut feels is intuition. Instead of logic, intuition is your perception of nonverbal information through images. If you are in the process of understanding a person, their outer trappings are insignificant, and it is only who the person is what counts. To reveal a richer story, intuition gives the power to distinguish beyond the obvious to tell a richer story.

You need to watch out for these checklists' cues of intuition:

Respect your gut feelings

Pay attention to voices of your gut, in particular when connecting with someone for the first time, an automatic rejoinder that happens out of impulse. Gut feelings are as a result of if you are tensed up or at ease. As a cardinal response, gut feelings occur in an instant. They are meters of your inner truth that relay to you if you should trust someone.

Goosebumps feelings

Pleasant, intuitive shivers are goosebumps, and they happen when something strikes a chord in us in connection with our resonance to individuals that inspire or move us. Also, goosebumps occur in the course of going through déjà-vu and when you have never met someone before but still recognize them.

Listen to sparkles of insight

During a conversation with people, you may be impressed by those who come quickly. Watch out and stay alert. Or else, you might fail to spot it. For most of us, this crucial awareness is lost because of the inclination to move onto the next idea.

Look for insightful empathy

This cue happens when you have a passionate type of empathy through the feelings of someone's real emotions and symptoms within your body. So, while reading people, take note whether you had pain on your back when it wasn't there before, or if you are upset or depressed following a mind-numbing conference. To determine if empathy is at play, get feedback.

Discern emotional power

The vibe we radiate and the remarkable demonstration of our energy are emotions. It is with an intuition that we procure these emotions. For some people, you will be happy to be around them because they enhance your vitality and mood. Others tend to be draining; get away from them is what you want. Though it is undetectable, you can feel this 'subtle energy' feet or inches from the body. It's called chi in Chinese medicine, an essential healthy vitality.

Be aware of the presence of people

Though not substantially similar to our behavior or words, the accustomed energy we discharge is when we sense the presence of the people. It is typical of a rain cloud or the sun that borders around our emotional atmosphere. In the process of reading people, take note of if you get attraction with their presence or retreating due to the willies you are getting.

Watch people's eyes

Humans' eyes convey compelling forces. As the eyes cast off an electromagnetic signal, according to studies, the brain does the same. When you watch people's eyes, you will know if they are tranquil, sexy, mean, angry, or caring. Also, you will have the ability to determine if a person wants intimacy in their eyes or their eyes can give signs that they are comfortable. Even in their eyes, you will know whether they appear to be hiding or guarded.

Observe the feel of a hug, handshake, or touch

Most of us shake emotional energy, similar to an electrical flow during physical contact. You can ask yourself if a hug or handshake feel comfortable, warm, or confident. Or if it is repulsive so much that you wish to withdraw. You can know the sign of anxiety with someone's hand clammy or limp to suggest being timid or non-committal.

Listen to the tone of laugh and voice

Our voice's volume and tone are capable of telling a lot about our emotions. Vibration is as a result of sound frequencies. Notice how people's pitch of voice affects you in the course of reading them. Envisage if the tone is snippy, abrasive, and whiny or if their tone feels soothing.

To read people can be hard sometimes. It takes practice and courage. However, once you are past that, you will gain a significant advantage. Not only will you survive, but you will also thrive in all your relationships with others. People will approach you. Opportunities will come to you. And some people will want to be like you.

Chapter 25
The Secrets To Taking Control Of Your Life

If you feel like your life happens without you, it doesn't have to stay this way. This will tie in everything we've talked about up until now to teach the reader how to take their power back and become the one who controls their life instead of it being the other way around.

In order to be in control of your life, you need to feel in control. There can be a sense of safety to feeling like other people make the decisions for you. For one, the pressure is off of you, and you are not responsible for when things go wrong. However, it comes at a price. You will have to watch others decide your fate. You will never achieve success if you do not stand in your own power.

There are things in life that you cannot control. Things that happen outside of you and the choices other people are going to make fall into this category. You will never have control over your life if you do not learn to recognize the difference between what you can and cannot have any impact on. The first thing you need to remember here is that the only thing you truly can control is your actions and your reactions to stimuli.

This is not to say you cannot be provoked or tempted to lose your temper. When someone says or does something rude to you, it will only be natural to feel anger. You will have fleeting thoughts of acting out in anger. However, if your thoughts become a reality and you actually follow through with your impulses, you must hold yourself accountable for it. Everything we do is a choice we make. No one can make anyone else react in a certain way. It will be an empowering moment when you realize you have the power not to react when someone provoked you.

People who have hacked their minds place themselves in a position of having control over their lives. People who do not control their lives put themselves in the passenger's seat. They allow others to decide what is going to happen in their life. Everything a person does, they are gaining something from it. When someone relinquishes control over their life, they are able to place the blame on others when something goes wrong. That is one of the most prominent benefits of playing the role of a victim. If they are let go from their job, it is because their boss had it out for them from the start. If they never accomplished a goal they wanted to, someone held them back (the person they are in a relationship with tends to be the one this particular bit of blame is placed on). They can still be in a foul mood by the evening about a minor rude gesture from this morning, such as being cut off on the way to work. The problem with having this type of mindset is that you will always be unhappy and feel unfulfilled. Even if you shift the blame onto others, you will still feel a sense of shame within yourself.

If you want something, you have to hold onto it with both hands. If you want a music career, you have to create music and put it up somewhere that people will hear it. If you want to write a book, you need to start putting words down onto a page. If you want to start a business, you will need to bring in customers and develop a sellable product. All of these endeavors will require a lot of time and effort. Anyone who is famous for achievements such as these put in such work and had to pick themselves back up after a lot of rejections.

You must learn how to deal with rejection in a healthy way instead of internalizing it if you ever hope to be successful. When you are developing your career, especially in the beginning, you will deal with a lot of rejection. Many young people today find themselves dejected because they are sending out job applications every day and either not getting any replies or being met with letters that tell them their application was declined. If these rejections are taken personally, the person is at risk for developing what is known as post-graduation depression. This means they are worried that they will never find employment or start their future. This anxiety causes them to want to avoid it. This means they will stop sending out applications or any other behaviors to seek employment. This is what happens when you interpret rejection as a personal failure instead of what it really is.

When you are rejected by a business or a person, you were just not a good fit for that particular situation. For example, if you ask someone out on a date and they turn you down, they are not trying to say you are undesirable. They are just not romantically interested in you. Your job application being declined does not mean you aren't hirable. There are only a certain number of people they can hire. They had to look through a lot of resumes and they saw someone whose credentials matched what type of person they were looking for. This time it was someone else. You will suffer indefinitely if you internalize rejection because it continues to happen to everyone throughout life.

This is something else not to lose sight of. Misfortunes happen to everyone. No one gets what they want all the time. It is an extremely unhealthy thought pattern to fall into, to start buying into the idea that everyone else is given everything while you are denied. This will cause a number of ill effects. For one, you will likely fall into a state of depression. You will also come to be resentful of others. You will spend a lot of time angry, which is not good for any aspect of your health.

While you need to chase the things you want, there is one caveat to this. There are things you will not obtain no matter how hard you want it, most often this comes in the form of unrequited love or trying to fit into a certain social group. It could also be when you are trying to convince a friend not to decide you know is a bad one, and you can foresee the consequences it will mean for them. None of these situations are ones you can change. This is because the power lies with the other person. In order to have a relationship, both people need to want it. If the other person does not love you back, it will never be real. If you don't have someone's approval, no amount of effort will gain it. It is actually an act of taking back your power to stop trying relentlessly to obtain the impossible.

All have us have looked at a friend or colleague and thought how easily success comes to them. They seem to ooze confidence and make the right decision, every time. Even if something does not go their way, they seem to take it in stride. Maybe they even say something like "Well, I can chalk that up to experience." They make a mental note of the event and how things went awry to be dissected later. What you do not see them doing is hurling negative thoughts onto themselves because this is a derailed that has nothing to do with attaining their goals.

It is important to visualize your goal to the point that you can really see yourself accomplishing it. You can consciously change your thought patterns to suit your path to success. "Don't sweat the small stuff" is a commonly known aphorism, but it can be hard to put into practice. To some, it does not come naturally where there is an inclination to overanalyze what one does and how one appears in his/her interactions.

Removing negative thoughts from your mind before they have a chance to take root can free up a great deal of space in your mind before they have a chance to impede you on your journey to prosperity. The concept is sort of like Disk Cleanup on your computer. You can focus on the positive and tidy up the space in your mind.

Along with your own negative thoughts, we all have situations that cause stress for us. Some of those are unavoidable like the line being especially long at the grocery store or the traffic being particularly congested when it rains. However, we should also explore ways we can remove frustrations from our lives that are a matter of choice.

Let's delve into some situations and, as we do, perhaps you can think of personal examples. First, there is a restaurant very near your house. Sometimes when you go, the experience is great; you have a good time and enjoy yourself. However, there is one particular waitress who is negative to the extent that her expression is constantly unhappy and even her voice shows little liveliness and vigor. When you eat your meal in this type of environment, it is bad for digestion and the mood lingers into your afternoon. You have a couple of options such as trying to ignore her, but instead, you can go to another restaurant down the street if you peer in the window and see her there. Another example could be that you are playing your favorite video game and someone is typing vile things on the screen and it is impacting your mood. Most games have a block player function so you can return to having the pleasurable distraction that your game was designed to be. Lastly, training your mind away from self-destructive behaviors such as looking at your ex-partner's social media is vital to your path to success. This is an exercise in futility and will create bad feelings that will contaminate your path to success.

You can literally train your brain to stop obsessing over things that will cause feelings that will you stress inhibiting you from favorable outcomes. When stopping one behavior, it's important to replace it with something else that is better for you. Writing down goals you want to achieve is an important step to making them a reality. The fact that you have begun to remove negative thoughts from your mind, staying away from situations that cause anxiety which will rip thoughts of your goals from your mind and moved on to positive thoughts will leave you energized. Your mind is clear and your thoughts are calm. Meditate on your goals and visualize yourself doing the steps that will take you to them. Is your goal to learn another language? Visualize yourself purchasing a book and signing up for a class. See yourself making flashcards for yourself so you can quiz yourself on your vocabulary words. You have questions and the instructor assigns some of them to you. Envision that you decide to do all the questions because you want to get more practice and learn more. As you think deeply about this goal, you can make decisions such as assigning yourself moments where you will find someone to practice with and/or begin to think in your new language, perhaps for an entire afternoon.

With your freed up "disk space," your brain is working faster, thinking ahead about your goals. You are giving yourself positive affirmations so you are no long defeating yourself before you even set out to accomplish your goals. You are making plans that will take you closer to your goals.

Chapter 26
How To Avoid Dark Persuasion And Stop Being Abused

We are indeed human at the end of the day. It is because of this very reason that we get to dwell allot on the opinion of others in everything that we do. We always desire and adore getting validation from others so that we can subconsciously decide whether or not we shall be depressed. In this age of the millennial, the norm has become to just brag about their wealth on social media. A lot of these bragging are often than not the truth. This ultimately leads to one having a loose relationship with reality. Self-deception of this type can dig deep into the human spicy, that a victim of these may one day wake up and realize that their perfect world is only existent within their maids. Depression will closely follow suit. The first step to attempting to defend yourself from persuasion and manipulation is confronting the situation and taking the stance of breaking off any illusions you may have. You will not be able to proceed normally with your life. You have to be wary of the fact that you are in control of your own choices. Then make the conscious choice of seeing things for what they are. That deal, which seems too good to be true, could actually be just that... too good to be true. The other thing you should follow is to definitely trust your instincts. There are times that a lie has been told to you in the most skilled way imaginable, that you will end up believing. But you can feel an imbalance on some instinctive level between what should be, what is, and then what is being projected onto you. There may be no physical signs to show that hey, something is wrong, but you feel something is wrong. The next important thing when you ask questions is to listen to the responses. This may sound somewhat unbelievable because you'll listen to the answers. The truth is that our self-disappointment can make us choose the answers we receive. We tell ourselves that we listen, but we only pay attention to the answers we want to hear rather than to the answers we receive. You may have broken the illusions around you, but some of you are still clinging to the comfort of those illusions. The pain of confronting the situation would prevent you from listening to the real answers to your questions. Actual listening requires a certain sense of detachment, but this time around not from reality. You have to get rid of your emotions. Your detachment from our emotions would lead you to the next step, which would logically process the new information. It can complicate situations more than they already are to act irrationally. It makes your exit strategy so much difficult to let all the emotions simmer and spring to the surface. When you face the truth, the irrational part of you may want you to let it all go hell. Your rightly justified anger can inspire you to take steps to calm your emotions in the short term. But you may come to regret these actions in the long term. I'm not saying that you should deny your emotions; I'm not saying that you do not act on these emotions. First deal with the situations and later deal with your emotions.

Act quickly

It's great that you have come to terms with the reality of things. But defence against these dark manipulative tactics entail so much more. While attempting to defend you from the claws of these manipulators, is often intense and exhilarating at first. This intensity of these emotions may cause one to slowly slide into denial. The more you delay in taking any action is usually what accelerates the onset of this denial, and when it happens, there are high chances that you might relapse and end up getting trapped in the same web. This can be avoided by taking action immediately you realise that someone is trying to manipulate you. This can present itself in the simplest of ways like when informing a close friend of some reality of the particular situation may be all that's needed so set in motion a series of events that will eventually lead to your freedom. You should know that the fabric of illusion is made from tougher material than glass after making the choice to act. The illusion could work its way back into your heart with your emotions in high gear by using fragments of your emotions to fix it. When a liar is caught in a lie, he or she may attempt to recruit others to enforce that lie when they feel that they are no longer holding you. A deceptive partner with whom you have recently broken things off would at this point try to use the other mutual relationships in your life to change your mind. If you want to get out of this unscathed, you will need both your logic and instincts. Although the truth of the situation is that when you discover that you've been lied to consistently, you become emotionally scarred, so the issue of leaving the situation unscathed becomes silent. Priority should be given, however, to take the route that allows you to leave this toxic situation without harming yourself further. You're all over the place emotionally. Rage, anger, hurt, and deception is the iceberg's tip. But logically, you need to think. Keep your head above the water and warn yourself.

Get help fast

When you're trapped by other people's manipulations, confusion is one of the emotions you'd experience. This helps cloud your rational thinking and leaves you feeling helpless. You might even question the reality of what you are facing at this point. It would lead to denial if you continue to entertain these doubts. You're probably going to want to conclude you've got the whole situation wrong. That you misunderstood some things and came to the wrong conclusion. Such thinking would drive back to the manipulator's arms. Resist the urge to give in by receiving a second opinion. People go to another doctor in a health crisis to get a second opinion. This is to remove any iota of doubt about the first diagnosis that you may have and to affirm the best treatment course for you.

Similarly, getting another person's opinion can help you discern the truth of the situation and what might be your next steps. Just remember, it's better to go to someone who has proved countless times they're interested in your best. The next step is to confront the perpetrator if you have the help you need. For this, I suggest you choose the scene or location. Choose a place you know that gives you the upper hand. On your part, that would require some careful planning. If the perpetrator exists in the cyber world, especially if the person swindled you of your money, you would have to involve the police and the relevant authorities. Do some of your own investigations so as to ascertain the truth. After you face the perpetrator and take the necessary steps to get out of the situation, you must start the healing process quickly.

The scale and gravity to which you were hurt, manipulated or abused do not matter. You must be able to walk past it and wait until you can "heal" your wounds, rather than sitting on your couch and reliving the past. If you don't do anything about it, an unhealthy scab could form over the wound, which would make you as vulnerable if not more than you had experienced. Speak to a counsellor, attend therapy, and take an active part in facilitating the healing process, whatever you choose to do. It won't happen overnight, but you are sure that you get closer to improving every day and every step you take in therapy.

Trust your instincts

While your brain interprets signals based on facts, logic, and sometimes experience, your heart works in the opposite direction by screening information through an emotional filter. The only thing that picks up vibrations is your gut instinct, which neither the heart nor the brain can pick on. And if you can groom to the point where you recognize your inner voice and are trained to react to it, you will lower your chances of being seduced by people trying to work on you with their manipulative will. To begin with, it's hard to recognize this voice. And that's because we allowed voices of doubt, self-discrimination as well as the critics ' loud voices within and without drowning out our authentic voice over the course of our lives. Your survival depends on this voice or instinct. So, trust that when it kicks in, your brain neurons can still process things in your immediate vicinity.

Some people call it intuition, and some refer to it as instinct, especially when it comes to relationships, they are undoubtedly the same thing. You must accept that it may not always make logical sense to start trusting your instincts. If you've ever been in the middle of doing something and experienced the feeling of being watched all of a sudden, then you know what I mean. You don't have eyes at the back of your head, there's no one else with you in the room, but you get the tiny shiver running down your spine and the "sudden knowledge" you're watching. That's what I'm talking about. The first step to connect with your instinct is to decode your mind with the voices you've let in. With meditation, you can do this. Forget the chatter of "he said, she said." Concentrate on your centre. You are the voice you know. Next, be careful about your thoughts. Don't just throw away the eclectic monologs in your head. Rather go with the thoughts flow.

Why do you think of a certain person in some way? How do you feel so deeply about this person, even if you only knew each other for a few days? What's that nagging feeling about this other person that you have? You get more tuned to your intuition as you explore your thoughts and understand when your instincts kick and how to react to it. You may need to learn to take a step back to pause and think if you are the kind of person who prefers to make spur decisions at the moment. This moment in which you pause gives you the opportunity to really reflect on your decisions and evaluate them. The next part is a hard part and it couldn't be followed by many people. Unfortunately, you can't skip or navigate around this step. This part has to do with trust. You need to be open to the idea of trusting yourself and trusting others to be able to trust your instinct. Your failure to trust others would just make you paranoid, and it's not your instincts that kick when you're paranoid.

Chapter 27
Applying Manipulation and Mind Reprogramming in Different Roles

As you are aware of by now, your thoughts play a major role in your choices and decision-making in everyday life. In this segment, we look at these different roles and the way negative thinking can hold us back. Your thoughts influence your behaviors, which in turn, shape your life. So, learning to regulate negative thinking and replacing it with positive paradigms is quintessential for becoming successful. Starting from your personality and your role, you will learn about simple exercises you can use to change your negative paradigms into positive ones.

Salesperson

What is the primary goal of a salesperson? To increase sales. To do this, you must be good at networking and effectively communicating with others. Apart from that, you must also be able to influence the decisions of others. To do this, you need to have not only an acute sense of self-awareness but also an awareness of other desires. If your goal is to increase your sales, then any thinking pattern that doesn't help you attain this goal is undesirable. Perhaps you doubt your ability to sell, or maybe you believe that the customer will never make a purchase. In such a situation, attaining your goal can become extremely tricky.

To become a great salesperson, you need plenty of self-confidence. Confidence must come from within, and unless you truly believe in yourself and what you are pitching, others will have difficulty believing you. Any negative beliefs you have about yourself will effectively hinder your selling skills. If you seem meek, unsure of yourself, and fumble while talking, this won't elicit confidence. Don't let your negative thought patterns hold you back.

Exercises

Your Circle

The company you keep matters. Past a certain age, who you are and how you interact with others is usually a direct representation of the kind of people you spend most of your time with. So, take some time and think about your different circles of friends. What do each of them represent? Are you surrounded by free-thinkers or followers? What are the different emotions brought by your colleagues and peers? How do the conversations usually? Are they filled with positivity or with unnecessary pessimism? If you believe you are surrounded by mostly negativity, then it is time to break free of this toxic energy you have voluntarily surrounded yourself. If someone doesn't add to your growth as a person, make peace with it, talk to them, and move on. You cannot grow when others around you hold you back. Instead, surround yourself with people who bring about change, roundedness, and a sense of purpose. Seek out those with ambition and want more for their life and others.

Dealing with Adversity

Whenever faced with a challenge, it can be easy to give up or blame others for what you might lack. If you want to be a successful salesperson, then it is better you start questioning yourself. Whenever you face a setback in life, try to analyze the situation and yourself. Every setback is a lesson life is trying to teach you. Unless you learn this lesson, you are bound to make the same mistakes again. Whenever you face adversity, ask yourself, "What is good in this challenge? Then, ask yourself, "What is the lesson I have to learn?" Character growth and development occur only when you manage to learn from your mistakes. Keep in mind that adversities are unavoidable in life whereas optimism is a choice. Whenever you are faced with adversity, learn to change your response.

Try to look for some humor in every situation. Dealing with a stressful situation increases the production of cortisol (stress-inducing hormone). The best way to diffuse such tension is by looking for some humor. When you learn to deal with a stressful situation using humor, it helps put all your challenges into a proper perspective.

Taking Control

If you find yourself thinking thoughts like: "I don't have a good enough marketing strategy, "My competition is better," or "My audience is lousy," or "My territory is no good," it merely reflects your inability to solve. If you believe that something external always guides your decisions and course of action, you can never be in control of the situation. If you want to be a successful salesperson, you need to understand that you are in control of your fate. Yes, there will be sometimes when it's beyond your reach, but these situations will only get the better as you try to improve in your work. If business seems slow, it could be time to change your approach or redefine your efforts. Try meeting with seasoned professionals who can act like mentors in your field that are willing to share their experiences and guide you to a new approach. Stay open to change and be prepared to put in the hard work. Keep learning and improve your skills to outsell your competition. Unless you take control of your mindset, you won't get ahead in this business.

Replace the negative thoughts with more positive ones like, "I can always learn and improve my marketing strategies," "If I work hard, I can improve myself," or "I can find a way to connect with my audience." A successful salesperson knows how to manage results and take control of the situation without making excuses.

Manager Dealing with Staff

As a manager, you must be able to not just convince your subordinates to stick with your plan of action but must be able to encourage and motivate them too. If you cannot do this, then you cannot be a successful manager. Any negative thought patterns you have could effectively hold you back and prevent you from successfully managing your team. Your success as a manager depends to a great extent on your ability to make the most of your team members. Even if you give your 100% but cannot make your subordinates contribute, then it is not going to get you anywhere.

If you think others will not listen to you or don't believe in your ideas, then it shows poorly on your managerial abilities. Taking a stand for yourself and voicing your opinions with confidence makes all the difference when it comes to leadership. If you don't have confidence in your leadership abilities, then you cannot lead anyone. If you think no one will listen to you, then it will become incredibly difficult to make others listen to you.

Exercises

Challenge your Thoughts

Negativity can creep in unannounced, and it can hit you like a ton of bricks. If your thoughts often start with words like "shouldn't," "will not," or "cannot," change the conversation. Whenever you start thinking this way, challenge the reasoning behind it.

Ask yourself whether your thoughts are relevant to this moment. Is this productive?

Once you have your answer, try to replace them with a positive vision. It might take a while, but eventually, you will be able to ignore negativity and instead concentrate on working towards a positive outcome.

An Employee Dealing With a Negative Boss

Dealing with bosses is never easy, and it is even more difficult if you deal with a difficult boss. If you want to excel in your chosen profession, learning to deal with any type of higher up is a very important skill. You might desire to be the star employee or get a job that enables you to make the most of your skillset. It is never easy working under someone unless you have a good rapport with the right person. You probably aspire to be more successful or do some meaningful work that will make others take notice of you. If you want to stand apart from the crowd, you must be able to hold your ground.

Dealing with negative thoughts and paradigms can be tiring. If you have convinced yourself that your boss doesn't like you, then you might lose interest in going to work or completing work on time. If you are riddled with self-doubt and regularly question every move you make or think you aren't suited for your job and responsibilities, you are self-sabotaging.

Exercises

Feed Your Mind

Be extremely cautious of what you feed your mind and soul. Fill yourself with positivity and allow this positivity to propel you towards a positive future. If you are dealing with a negative workspace or dealing with negative paradigms, find a way to change the conversation.

There are different ways to do this. You can use positive affirmations or even practice simple meditation. Whenever you feel negativity creeping in, take a break from your work, and concentrate on what you need at that moment to make yourself feel better. Before you accept a thought, question it. Don't blindly accept everything that you think to be the truth.

Limit Interactions

Negativity can spread very quickly if you are not careful. To prevent this, limit your personal exchanges with the negative boss as much as you can. Don't allow them to get into your mind. If someone is affecting you, regardless of their position or yours, you have already given them permission to do so. No one can make you feel bad unless you let them. You do this, unknowingly. Therefore, make a conscious effort to keep this negative person from entering your thought patterns.

Loner

It's never pleasant to feel like a social outcast. Maybe it makes you uncomfortable to socialize with strangers or anyone at all, and in turn, others feel uncomfortable around you. So, you might think it's easier to be alone. When you do get a chance to socialize with others, it becomes incredibly difficult for you to start a conversation or make worthwhile connections. If you feel others don't like you or that you aren't interesting enough to hold a conversation, it will harm your self-confidence, self-worth, and self-esteem.

Chapter 28
Mastering Your Emotions

What is an emotion? Emotion is someone's personality that usually consists of their feelings, as opposed to their thoughts. A conscious intellectual reaction such as happiness or sadness is experienced as a strong feeling often directed toward a specific object and typically accompanied by physiological and behavioral changes in the body.

Emotion is a determining factor that affects how we live and interact with others. To most, it may look like their feelings rule humans. There are many different types of emotions. They influence the choices people make, the actions taken, and the perceptions we have are all affected by the feelings we are experiencing at any given moment. Some of the basic emotions experienced are;

Fear

Fear usually is a response to impending danger. In the animal kingdom fear is a mechanism of survival. It may be a mild caution or extreme paranoia. Fear is a powerful emotion that can also play an essential role in survival. Someone who is facing some danger or experiencing anxiety goes through flight response. The muscles become tense, the heart rate and respiration increase, and the mind become more alert, priming the person's body to either run from the danger or stand and fight. This reaction of fear helps to ensure that one is prepared to deal with threats in his surroundings effectively.

The body language of a person experiencing fear can include:

- Facial expressions like broadening the eyes
- Efforts to flee from the danger and hide from threats
- Heavy and rapid breathing
- Heartbeat rate increases

People react to fear in different ways. Most people might be more sensitive to worry, and specific situations or objects might be more likely to trigger the fear. Some people, instead, may develop a more similar reaction to the anticipated threats or even the thoughts about the potential dangers. This act is what is believed to be anxiety. Social anxiety includes an anticipated fear of social situations. While others may look for fear-provoking positions. Extreme sports and various other thrills may be fear-inducing, but for some people, it might seem to thrive and even tend to enjoy such feelings.

The act of being exposed repeatedly to a situation of fear may lead to some sort of acclimation and familiarity, which may reduce feelings of anxiety and fear. It is the main concept behind exposure therapy. Therapists use this technique to gradually expose their patients to the things which frighten them in a safe and controlled manner. Ultimately, the feelings of fear may begin to decrease.

Surprise

The surprise is normally quite brief and is normally distinguished by a physiological surprise response after something unexpected. Shock can be either positive, negative, or neutral. A great example of a pleasant surprise may be arriving at your workstation only to find that your colleagues have gathered together to celebrate your birthday.

The body language of someone who is surprised involves:

- Verbal reactions such as yelling, breathless or shouting

- Facial expressions like opening the mouth or widening the eyes,

This type of emotion can easily trigger a flight response. When surprised, people might be subjected to a burst of adrenaline which helps prepare one's body to either flee or fight. This type of emotion may have an impact on human behavior. For instance, studies have indicated that individuals tend to notice unexpected events suspiciously. It is the main reason why in the news, unusual and extraordinary games will always stand out in memory in comparison to others. Studies have also found that people are normally more swayed by surprising and unexpected arguments.

Disgust

The body language expressed during this kind of emotion involves:

- Facial gesture such as curling the upper lip

- Folding of the nose

- Turning away

- The body also reacts physically such as vomiting or retching

The sense of disgust may originate from several things, including an unpleasant sight, taste or smell. Disgust is a normal reaction experienced when individuals taste or smell foods that are rotten or gone bad. Infection, poor hygiene, rot, death, and blood may also trigger a similar response. People may also be subjected to moral disgust when they take not of other people engaging in behaviors which they consider immoral, evil or distasteful.

Happiness

This emotion is the one people tend to strive to achieve. The number of researches done on happiness has dramatically increased since the 1950s. A field that has researched the state of happiness is positive psychology, a branch of psychology.

Happiness can sometimes be expressed through:

- A relaxed body language

- Facial expressions like laughing, playfulness, and smiling

- An upbeat and pleasant tone

While joy may be considered to be one of the basic human emotions, the things that we think may create happiness tend to be influenced heavily by culture. The realities of what exactly contributes to happiness tend to be much more highly individualized and complex. Studies have come out to support the idea that happiness and health are connected. It shows that joy may influence both the mental and physical health. Happiness linked to a variety of outcomes, including increased longevity and increased marital satisfaction. Equally, someone who is unhappy is likely to have a range of poor health outcomes. Stress, anxiety, and depression have all been linked to a number of things like lowered immunity, decreased life expectancy and increased inflammation.

How to Master Your Emotions

"Anyone may become angry and that is easy. However, to be angry with the right person at the right time, and for the right purpose and in the right way – that is not within everyone's power, and that is not easy." This is a quotation made by Aristotle during 350 BC.

Mastering your emotions will require a degree of awareness which not everyone is intending to have. Understand various emotional languages and also become familiar with their shortcomings. Once you get a little bit discerning on how to effectively express your emotions, then the more you will get more knowledgeable at understanding them. There are numerous ways someone can learn to master emotions and help himself not to let emotions dictate behavior.

Cultivate Emotional Intelligence

Emotional Intelligence might suggest different images. It simply refers to the process of getting your brain to create the most useful instance of the essential emotional concepts in any given situation. This act will require you to adjust your concepts: rather than piling all similar emotions under one universal term. You can try to learn more about the nuanced meanings of a variety of emotions like misery which may come in a variety of flavors such as enragement, bitterness, mortified and irritated. There are a number of ways to make yourself feel great, like being thrilled, ecstatic, and grateful.

From the moment you wake up each morning, you have a choice to either enjoy the day, be fully engaged in a life or be motivated. The best way to break this cycle of negative emotions is by adopting higher positivity. Actively looking for new ways to use positive emotions to adapt to negative moods will help us repair our spirit and also improve our responses and overall thinking. Emotions are the building blocks of who we are, and also, they are an innate potential meant for our survival. We all can experience negative and positive emotions, and also learn how to regulate these feelings to come up with the best response, which is the key to our emotional intelligence.

Getting to know more about the skill to differentiate between the subtle nuances of different emotions will make you an emotion specialist and also assist your brain by giving you more options to forecast and categorize different sensations more efficiently. It will also help you tailor your actions better to your environment

Learn to Say NO

It involves people who have codependency issues. Most people with codependency symptoms have a lot of trouble saying NO and sticking with it. They might be able to say NO a couple of times, maybe once or twice, but after some more urging, they give in. Salespeople love dealing with people with this symptom. It is because they know that they can get them to buy what they offer at a price that gives them a fat commission.

Personal relationships are also affected by codependents because they become a pushover. One controlled in an emotional state where his/her needs usually not considered. You think you're a loving partner, but you're damaging the relationship further. This because you aren't honest with yourself and your partner.

In the world of business, one needs to learn to say no. In industry, learning to say no is one of the most powerful negotiating tools.

Collect New Experiences

Instead of reiterating the same old patterns and behaviors over and over again, then let go of the past and begin accumulating new experiences. Be a collector of experiences. Unique experiences that you collect through reading books, learning about other cultures, going on trips and adventures, watching movies, acquiring new perspectives, trying fresh foods, learning new words and offering new opportunities to constructing your experience in a variety of ways. Doing all these things will assist you to shift your perspective and also what you expect is going to happen next. Getting to acquire new skills helps in encouraging your brain to create new concepts and also bind the old ones in new ways. Thereby, affecting future behaviors and predictions. By placing yourself in other people's shoes, you will be able to develop and strengthen your compassion muscle. For example, enlarging your vocabulary may lead to improved emotional health by offering new concepts. This, in turn, will not only assist you to become better furnished to deal with a number of circumstances but also improve your negotiation skills and increase your empathy?

Recognize Unhelpful Thought Patterns

People's behaviors usually are a response to emotions, and these emotions tend to be a response to one's thoughts. Most people carry around what cognitive behavioral therapy is known as cognitive distortions. It can further explain as someone filtering out the experiences to demonstrate some point about their core beliefs. If one feels unworthy of unconditional love, or somewhat destined for a life that's filled with misery, the person will progress by accumulating the evidence to show support for that. We tend to catastrophize minor issues, personalizing things which had nothing to do with us, and in the process limiting our potential. This is where a fact-checking majority of these rampant thoughts might be helpful. We should understand that if our thought patterns are resultant of correct facts or just assumptions that we are making. When we can engage more efficient with all of these thoughts, then the subsequent negative emotions, which follow may get suffocated. Proceed by distinguishing your emotions more finely. Learning to identify different emotions with greater granularity will help us in regulating our emotions much better.

This is because the emotions provide individuals with more information on how to regulate their behavior and also on how to deal with various circumstances. Research shows that individuals who are able to separate ideally between the emotions stood a less chance of resorting to feeling overwhelmed especially under stress. In another study, it was established that when individuals who fear snakes labeled their emotions using different fear and anxiety words. They eventually became less anxious when they were around snakes. Moreover, when the 6th and 5th graders decided to enrich their vocabulary with emotion words, they were able to perform much better in class and also improve their social behavior when in school. On the other hand, people with depressive disorders and social anxiety tend to portray and experience fewer negative emotions in their day to day lives.

Conclusion

Now that you have tapped into the forbidden truth that makes Dark Psychology possible, you have the tools and know the techniques necessary to know what people are thinking — and even change their thoughts.

You can't allow yourself to be discouraged when these techniques don't work out when you try using them in the beginning. Always keep in mind that like any skill, NLP takes practice. Like anything worth learning, you must fail at it sometimes so you can learn from those mistakes. Your only failure would be no longer trying. So, practice, practice, practice!

Remember that deception is not always practiced on other people. We can often self-deceive to preserve our self-esteem. Telling ourselves that we can achieve certain goals when all the evidence points to the fact that we can't is a healthy form of deception, but self-deception can lead to serious delusions.

Whatever happens in the novice stages of your path to becoming a master of manipulation and persuasion, you must remember your end goal. Ask yourself in the beginning why you want to do this and keep coming back to that when it gets hard. Never give up; you are to master these skills.

I hope that through this book, you have realized that brainwashing, manipulation and persuasion depends greatly on an authoritative command of words. You might be able to list twenty manipulation techniques from memory; you may be able to get someone with little psychic resistance to go with your ideas.

You may have gotten to the end of the book — and you may have all the knowledge necessary to manipulate people — but you are just beginning when it comes to putting this all into practice.

In fact, experts say that forming long-term memories requires us to get it into our brains in multiple different ways. Reading is just one way you have learned about manipulation and persuasion. You have several options for the next way you learn it.

Maybe you will keep a journal of different people you have tried these techniques on (if you do that, keep this journal somewhere private, of course). You can try re-reading our book but taking notes along the way so you can get the information into your writing hand. They say the best way to learn is to teach, so you can try to talk with your friends about what you have learned. This will also communicate to them that you did not intend to use these techniques on them, so there is an added bonus.

We want to think you sincerely for taking the time to read through to the end. Hopefully, it contained everything you needed.

www.ingramcontent.com/pod-product-compliance
Lightning Source LLC
Chambersburg PA
CBHW071356210526
45465CB00001B/117